Biological Networks

Biological Networks

Editor: Carol Connolly

www.callistoreference.com

Callisto Reference,
118-35 Queens Blvd., Suite 400,
Forest Hills, NY 11375, USA

Visit us on the World Wide Web at:
www.callistoreference.com

ISBN: 978-1-64116-355-2 (Hardback)

Cataloging-in-Publication Data

Biological networks / edited by Carol Connolly.
    p. cm.
Includes bibliographical references and index.
ISBN 978-1-64116-355-2
1. Bioinformatics. 2. Computational biology. 3. Biology--Data processing.
4. Systems biology. I. Connolly, Carol.
QH324.2 .B56 2020
570.285--dc23

Table of Contents

Preface

Any system with sub-units linked into a whole is known as a network. A biological network is a network of biological systems. It provides a mathematical representation of the connections between ecological, evolutionary and physiological studies. The prominent networks in biology are metabolic networks, food webs and gene regulatory networks. Biochemical reactions connect the chemical compounds of a living cell and convert one compound into another. This biochemical network of reactions is known as metabolic network. All organisms are connected to each other through feeding interactions and form an intricate web which is known as food web. All cells have complex gene regulatory networks in which the gene activity is regulated by transcription factors. The topics included in this book on biological networks are of utmost significance and bound to provide incredible insights to readers. It strives to provide a fair idea about this discipline and to help develop a better understanding of the latest advances within this field. For all readers who are interested in biological networks, the case studies included in this book will serve as an excellent guide to develop a comprehensive understanding.

This book is a comprehensive compilation of works of different researchers from varied parts of the world. It includes valuable experiences of the researchers with the sole objective of providing the readers (learners) with a proper knowledge of the concerned field. This book will be beneficial in evoking inspiration and enhancing the knowledge of the interested readers.

In the end, I would like to extend my heartiest thanks to the authors who worked with great determination on their chapters. I also appreciate the publisher's support in the course of the book. I would also like to deeply acknowledge my family who stood by me as a source of inspiration during the project.

Editor

3D structure prediction of histone acetyltransferase (HAC) proteins of the p300/CBP family and their interactome in *Arabidopsis thaliana*

Amar Ćemanović[1], **Jasmin Šutković**[1], **Rabab Elamawi**[2], **Waleed Elkhoby**[2], **Mohamed Ragab Abdel Gawwad**[1]

[1]Genetics and Bioengineering department, International University of Sarajevo, Ilidza, 71220 Bosnia and Herzegovina
[2]Rice Research and Training Center, Sakha, Egypt
E-mail: mragab@ius.edu.ba,mrgawad@hotmail.com

Abstract

Histone acetylation is an important posttranslational modification correlated with gene activation. In *Arabidopsis thaliana* the histone acetyltransferase (HAC) proteins of the CBP family are homologous to animal p300/CREB (cAMP-responsive element-binding proteins, which are important histone acetyltransferases participating in many physiological processes, including proliferation, differentiation, and apoptosis. In this study the 3-D structure of all HAC protein subunits in *Arabidopsis thaliana*: HAC1, HAC2, HAC4, HAC5 and HAC12 is predicted by homology modeling and confirmed by Ramachandran plot analysis. The amino acid sequences HAC family members are highly similar to the sequences of the homologous human p300/CREB protein. Conservation of p300/CBP domains among the HAC proteins was examined further by sequence alignment and pattern search. The domains of p300/CBP required for the HAC function, such as PHD, TAZ and ZZ domains, are conserved in all HAC proteins. Interactome analysis revealed that HAC1, HAC5 and HAC12 proteins interact with S-adenosylmethionine-dependent methyltransferase domain-containing protein that shows methyltransferase activity, suggesting an additional function of the HAC proteins. Additionally, HAC5 has a strong interaction value for the putative c-myb-like transcription factor MYB3R-4, which suggests that it also may have a function in regulation of DNA replication.

Keywords *Arabidopsis thaliana*; histone acetyltransferase; 3D structure; interactome; functional annotation.

1 Introduction

In order for the whole genome to be packed into the nucleus of a eukaryotic cell, the DNA is associated with special proteins, called histones, to form chromatin fibers. Histones are small proteins that function in DNA packaging and gene regulation (Hardin et al., 2012). The region of DNA that is part of the core particle is inaccessible to transcription factors and therefore transcriptionally inactive. Various studies demonstrate the direct association between modification of chromatin structure and levels of gene expression (Allfrey et al., 1964; Katsani et al., 2003; Martinowich et al., 2003; McGraw et al., 2007; Kishimoto et al., 2006; Hublitz et al., 2009). These modifications can be in the form of methylation, ubiquitination, phosphorylation, ADP-ribosylation and acetylation. Of those mechanisms, the best known is acetylation – modification by the action of histone acetyltransferase (HAT) enzymes (Sterner and Berger, 2000).

HAT proteins function by transferring acetyl groups from acetyl-CoA to lysine residues on the N-terminal tails of histones. Thereby, they lower the affinity of histones for DNA and, accordingly, make the DNA more accessible to the transcription machinery. The process is reversible, with the deacetylation being performed by deacetylase enzymes. Based on sequence homology, histone acetyltransferases have been organized into four families: GNAT, MYST, p300/CBP and TAFII250. The p300/CBP family is especially important since it acetylates not only histones, but also various other proteins, such as the cAMP response element-binding (CREB) protein, p53, HIV-1 Tat protein and Stat3. Thus, the proteins of this family are also termed factor acetyltransferases (Loidl, 1994).

The p300 / CBP family encompasses the p300 and CBP proteins in animals and homologous proteins in other eukaryotes. Due to their high homology and similarity in structure and function, p300 and CBP have until recently been treated as one protein. Since it suits the purpose, this paper does likewise. p300 / CBP is a large, universally expressed, regulatory protein found in many multicellular organisms. It is involved in a wide variety of cellular processes, including differentiation, cell cycle control and apoptosis (Loidl, 1994). In addition, it is shown to participate in tumor-suppressor pathways, cell transformation and action of cellular oncogenes, associated with viral oncoproteins, contribute to the differentiation of specific cell lineages (processes of hematopoiesis and myogenesis), play a role in development and participate in transcriptional repression pathways (Goodman and Smolik, 2000). It is also suggested that p300/CBP is brought in contact with specific promoters by physical interactions with sequence-specific transcription factors (Vandel and Trouche, 2001).

On the molecular level, besides CREB, p300/CBP interacts with numerous promoter-binding trascription factors including nuclear hormone receptors and activators such as c-Jun, c-Fos, c-Myb etc (Sterner and Berger, 2000). Defects in its expression have been linked to a large number of developmental abnormalities, such as Rubinstein-Taybi syndrome (RTS) in humans (Goodman and Smolik, 2000). Additionaly, Vandel and Trouche (2001) have shown that p300/CBP physically associates with a histone methyltransferase enzyme.

Many conserved domains are found in the p300/CBP protein, including the bromodomain, CREB-binding domain (KIX), N-terminal nuclear receptor-interacting domain (RID),glutamine/proline-rich domain (QP), three cysteine/histidine-rich regions (CH1, CH2 and CH3) and a bipartite nulear localization signal (NLS-BP). The CH regions contain four zinc finger motifs: TAZ1, TAZ2, ZZ and PHD (Yuan and Giordano, 2002). The domains RID, KIX, TAZ1 and TAZ2 function in binding to transcriptional activators and regulators (Goodman and Smolik, 2000). The HAT function is performed by a large conserved region that spans from the PHD domain to the ZZ motif,with a putative CoA-binding domain being located in the middle (Ogryzko et al., 1996).

Although the human p300/CBP is well characterized and has its crystal structureexperimentally determined (Liu et al., 2008), little is known about its orthologs in Arabidopsis thaliana. The p300/CBP family in Arabidopsis includes the proteins: HAC1, HAC2, HAC4, HAC5 and HAC12 14, all displaying a significant degree of homology to the p300/CBP protein (Deng et al., 2007). The members of this family have previously been called PCAT proteins (p300/CBP acetyltransferase-related proteins) (Bordoli, 2001).

All five members of the family function in the nucleus, with HAC4 additionally being found in the chloroplast. Out of them, only HAC2 has been shown to lack direct HAT activity (Bordoli, 2001). HAC1,

HAC2 and HAC12 also display transcription cofactor activity, while all members of the family exhibit zinc ion binding activity. Furthermore, HAC1, HAC5 and HAC12 have been shown to be involved in the process of flower development (The UniProt Consortium, 2012).

All HAC proteins contain ZZ and TAZ zinc finger domains, implicated in interactions with transcription factors, and a C-terminal HAT domain that provides HAT activity in vitro. As compared to the p300/CBP protein, the HAC proteins differ in four main aspects: 1. they lack a bromo-domain; 2. they lack a KIX domain; 3. they have two ZZ zinc finger domains, whereas p300/CBP has only one and; 4. they lack the QP region near the C-terminus (Deng et al., 2007).

This in silico research is aimed to attribute additional functions to the HAC proteins of the p300/CBP protein family in *Arabidopsis thaliana* by interactome analysis. In addition, the phylogenetic relationship of the proteins will be assessed using multiple sequence alignment and phylogenetic tree construction, and the subcellular localization of each HAC protein determined. Finally, the three-dimensional (3D) structures will be predicted and confirmed with Ramachandran plot analysis and several validation tools available at the Swiss Model server.

2 Materials and Methods

The sequences for the HAC proteins of the p300/CBP family (HAC1, HAC2, HAC4, HAC5 and HAC12) were obtained from National Center for Biotechnology Information (NCBI) database (Sayers, 2009). Additionally, TAIR (The Arabidopsis Information Resource) ID numbers were obtained (Lamesch et al., 2012). The NCBI and TAIR accession numbers are shown in Table 1.

Table 1 HAC proteins of the p300/CBP family and their Ids.

HAC protein	Gene ID code	TAIR IDs
HAC1	NP_565197.3	At1g79000
HAC2	NP_564891.4	At1g67220
HAC4	NP_564706.1	At1g55970
HAC5	NP_187904.1	At3g12980
HAC12	NP_173115.1	At1g16710

Multiple sequence alignment (MSA) has been performed using the ClustalW software (version 2) located on the website of the European Bioinformatics Institute (EBI), using default options (Larkin et al., 2002; Goujon, 2010). MSA is an invaluable bioinformatics tool used to measure the similarity between sequences, examine patterns of conservation and variability and derive evolutionary relationships (Lesk, 2002). ClustalW uses a progressive method of alignment, meaning that it aligns the sequences one by one, instead of aligning all at once (Claverie and Notredame, 2007).

In order to infer the evolutionary relationship between the HAC proteins, a phylogenetic tree was constructed using Phylogeny.fr, a web service for phylogenetic analysis of molecular sequences (Dereeper et al., 2008). The service was run on default settings, and the steps that it performed to construct the phylogenetic tree involved multiple sequence alignment, alignment organization and construction and visualization of the phylogenetic tree using different integrated tools (Castresana, 2000; Chevenet et al., 2003; Guindon and Gascuel, 2003; Edgar, 2004; Anisimova and Gascuel, 2006; Dereeper, 2010).

As there exists no experimental published data on the 3D structure of the HAC proteins in Arabidopsis in the Protein Data Bank (Bernstein, 1977), the structures were hereby predicted with the help of the Phyre2 protein homology modeling server (Kelley and Sternberg, 2009). Phyre2 is a web-based service for protein structure prediction that is free for non-commercial use. Phyre2 has been designed and funded by the Biotechnology and Biological Sciences Research Council (BBSRC) from United Kingdom. Recently, a comparative structural study for the Arabidopsis HAC1 protein was performed to predict the most accurate structure of HAC1, where the HAC1 structural modeling by the Phyre2 server proved to be the most accurate, along with the Swiss Model homology modeling server (Cemanovic et al., 2014). A practical and widely cited molecular visualization tool, PyMOL, was used for structure visualization and representation (PyMOL).

After the generation of the 3D structures, structural evaluation and stereochemical analysis was performed using different evaluation and validation tools. Backbone conformation of all models and their stereo-chemical quality was evaluated by analysis of Ramachandran plots using the PROCHECK software (Laskowski, 1993).

Furthermore, 3D structure validation for all protein members was performed by the Swiss Model server with several incorporated validation tools (Schwede, 2003). The Swiss Model server includes a Swiss model workspace, an automated comparative protein modeling environment that incorporates several structural validation tools for model quality estimation. One of these tools is QMEAN6, a composite scoring function that is used to estimate the quality of a 3D protein model and to compare and rank alternative models of a target. The scores are given in values between 0 and 1, with higher values for better models (Benkert et al., 2008).

Another tool integrated into the Swiss Model Server is Verify3D, which assesses protein structures using three-dimensional profiles. It analyzes the compatibility of a 3D model with its own amino acid sequence (1D). The scores range from -1 (bad score) to +1 (good score) (Eisenberg et al., 1997).

The identification of domains in the five HAC proteins was performed using the online tool SMART (Simple Modular Architecture Research Tool) located on the website of the European Molecular Biology Laboratory (EMBL) (Schultz, 1998; Letunic et al., 2011).

The interactome of the HAC proteins was obtained with help of the Arabidopsis Interaction Viewer, which uses a database of 70944 predicted and 28556 confirmed Arabidopsis interacting proteins (Geisler-Lee, 2007).

The subcellular localization of each HAC protein was determined using WoLF PSORT, an online tool which functions by detecting known sorting signals and specific amino acid content in proteins. Experiments showed that the overall accuracy of WoLF PSORT is over 80% (Horton et al., 2007).

3 Results

3.1 Predicted 3D structure models and Ramachandran plot confirmations

For additional 3D structure validation and confirmation, statistical analysis with QMEAN6 and Verify3D software was performed. The results are shown in Table 2 and include also the Ramachandran plot PROCHECK results (Fig. 1-2).

3.2 Multiple sequence alignment and phylogenetic tree construction

The amino acid sequences of all HAC proteins is obtained from the NCBI database, analyzed in the ClustalW online tool for multiple sequence alignment. The obtained results are summarized in Table 3. The scores presented in the table are a measure of the similarity between the respective sequences, as determined by their pairwise alignment. A higher score represents a higher similarity.

The phylogenetic tree created for the HAC proteins shows that HAC1 and HAC12 as well as HAC4 and HAC5 are evolutionary close relatives. HAC2 is phylogenetically the most distant member of the family, diverging very early in the evolution of the HAC proteins (Fig. 3).

Table 2 QMEAN6, Verify3D and PROCHECK structure validation results.

HAC PROTEIN MEMBERS	VALIDATION TOOLS		
	QMEAN6 (0-1)	Verify3D	PROCHECK– favored region in %
HAC1	0.653	0.72	91.2
HAC2	0.630	0.69	91.4
HAC4	0.608	0.77	92.1
HAC5	0.649	0.82	91.2
HAC12	0.642	0.69	90.6

Table 3 ClustalW multiple sequence alignment scores for HAC proteins.

Seq. 1	Name	Length	Seq. 2	Name	Length	Score
1	HAC1	1697	2	HAC2	1367	26
1	HAC1	1697	3	HAC4	1456	44
1	HAC1	1697	4	HAC5	1670	50
1	HAC1	1697	5	HAC12	1706	75
2	HAC2	1367	3	HAC4	1456	24
2	HAC2	1367	4	HAC5	1670	24
2	HAC2	1367	5	HAC12	1706	26
3	HAC4	1456	4	HAC5	1670	53
3	HAC4	1456	5	HAC12	1706	44
4	HAC5	1670	5	HAC12	1706	48

3.3 Domain search

The domain search using SMART showed a high degree of similarity in domain presence and location among the HAC proteins, similar to what has been shown by recent research.

All the HAC protein members contain the following domains: zinc finger TAZ (transcriptional adaptor zinc-binding) domain, consisting of a distinct fold and is characteristic for proteins of the CPB/P300 family. In addition, HAC proteins in Arabidopsis have zinc finger ZZ domains, named so because they are able to bind two zinc ions, and a PHD (plant homeodomain) domain. All the domain locations are shown in Table 4.

3.4 Subcellular localization

The analysis of the subcellular localization using the WoLF PSORT tool showed that all HAC proteins are localized in the nucleus, which is also confirmed by TAIR. HAC2, HAC4, HAC5 and HAC12 were predicted to be localized in the chloroplast as well, whereas HAC1 was predicted also in the cytoplasm. Additionally, HAC2 and HAC4 were predicted in the mitochondria and plastid, respectively, although with low accuracy. The results are summarized in Table 5.

3.5 Interactome analysis

The results obtained using the Arabidopsis Interactions Viewer showed that HAC1, HAC5 and HAC12 interact with various proteins which are localized mainly in the nucleus. However, it provided no interactome for HAC2 and HAC4.

Table 4 Position of domains in HAC proteins.

PROTEINS	DOMAINS									
	TAZ 1		TAZ2		ZZ		ZZ 2		PHD	
	Start	End	Start	End	Start	End	Start	End	Start	End
HAC1	631	709	1582	1660	1398	1442	1518	1564	991	1064
HAC2	-	-	1275	1357	1093	1139	121	1265	704	763
HAC4	417	495	1345	1420	1160	1204	1280	1325	792	839
HAC5	612	690	1557	1632	1378	1421	1498	1542	972	1044
HAC12	638	716	1591	1669	1407	1451	1407	1451	1000	1073

Table 5 Results of localization analysis for HAC proteins.

Protein	Subcellular localization	Prediction accuracy
HAC1	Nucleus	12.5
	Cytoplasm	7.0
HAC2	Nucleus	9.0
	Chloroplast	3.0
	Mitochondria	1.0
HAC4	Chloroplast	6.0
	Nucleus	6.0
	Plastid	1.0
HAC5	Nucleus	7.0
	Chloroplast	6.0
HAC12	Nucleus	10.0
	Chloroplast	4.0

Common interactome partners for HAC1, HAC5 and HAC12 are 'smad nuclear-interacting protein 1' (At3g20550), 'S-adenosylmethionine-dependent methyltransferase domain-containing protein' (At1g45231) and 'transcription regulator/ zinc ion binding protein' (At3g47610). Interactome partners common just to HAC1 and HAC5 include 'N-terminal asparagine amidohydrolase family protein' (At2g44420) and 'DNA repair protein RAD51-like 1' (At5g20850). HAC5 and HAC12 show interaction with the 'DNA-directed RNA polymerase II subunit RPB1' (At4g35800), whereas HAC1 and HAC12 interacts with the 'abscisic acid-insensitive 5-like protein 5 (At1g45249). In addition, HAC5 shows strong interaction with both 'histone acetyltransferase GCN5' (At3g54610) and 'putative c-myb-like transcription factor 3r-4' (At5g11510). The results are summarized in Fig. 4 and Table 5. Proteins that interact with more than one HAC protein are written in the same color for easier recognition.

Table 6 Interactome of HAC proteins showing only significant interactions.

HAC proteins	Interactome	Interolog Confidence Value	Interolog Confidence	Interactome Location	Interactome function
HAC1	At1g10390	4	Medium	nuclear membrane, nucleus	**Nucleoporin autopeptidase**
	At1g45231	4	Medium	nucleus, cytoplasm	**S-adenosylmethionine-dependent methyltransferase domain-containing protein**
	At1g45249	4	Medium	nucleus	**Abscisic acid-insensitive 5-like protein 5**
	At3g20550	4	Medium	chloroplast, cytoplasm, nucleus	**Smad nuclear-interacting protein 1**
	At3g47610	9	Medium	nucleus	**Transcription regulator/ zinc ion binding protein**
	At2g44420	2	Medium	chloroplast, nucleus, extracellular	**N-terminal asparagine amidohydrolase family protein**
	At5g20850	2	Medium	nucleus	**DNA repair protein RAD51-like 1**
	At5g26680	4	Medium	nucleus, cytoplasm	**Flap endonuclease-1**
HAC5	At1g45231	8	Medium	nucleus, cytoplasm	**S-adenosylmethionine-dependent methyltransferase domain-containing protein**
	At3g20550	1	Low	chloroplast, cytoplasm, nucleus	**Smad nuclear-interacting protein 1**
	At3g47610	1	Low	nucleus	**Transcription regulator/ zinc ion binding protein**
	At4g35800	4	Medium	chloroplast, nucleus, plasmodesma, vacuole	**DNA-directed RNA polymerase II subunit RPB1**
	At2g44420	1	Low	chloroplast, nucleus, extracellular	**N-terminal asparagine amidohydrolase family protein**
	At5g20850	1	Low	nucleus	**DNA repair protein RAD51-like 1**
	At3g54610	12	High	nucleus	**Histone acetyltransferase GCN5**
	At5g11510	96	High	nucleus	**AtMYB3R4_putative c-myb-like transcription factor 3r-4**
HAC12	At1g10390	4	Medium	nuclear membrane, nucleus	**Nucleoporin autopeptidase**
	At1g45231	4	Medium	nucleus, cytoplasm	**S-adenosylmethionine-dependent methyltransferase domain-containing protein**
	At1g45249	1	Low	nucleus	**Abscisic acid-insensitive 5-like protein 5**
	At3g20550	4	Medium	chloroplast, cytoplasm, nucleus	**Smad nuclear-interacting protein 1**
	At3g47610	4	Medium	nucleus	**Transcription regulator/ zinc ion binding protein**
	At4g35800	4	Medium	chloroplast, nucleus, plasmodesma, vacuole	**DNA-directed RNA polymerase II subunit RPB1**

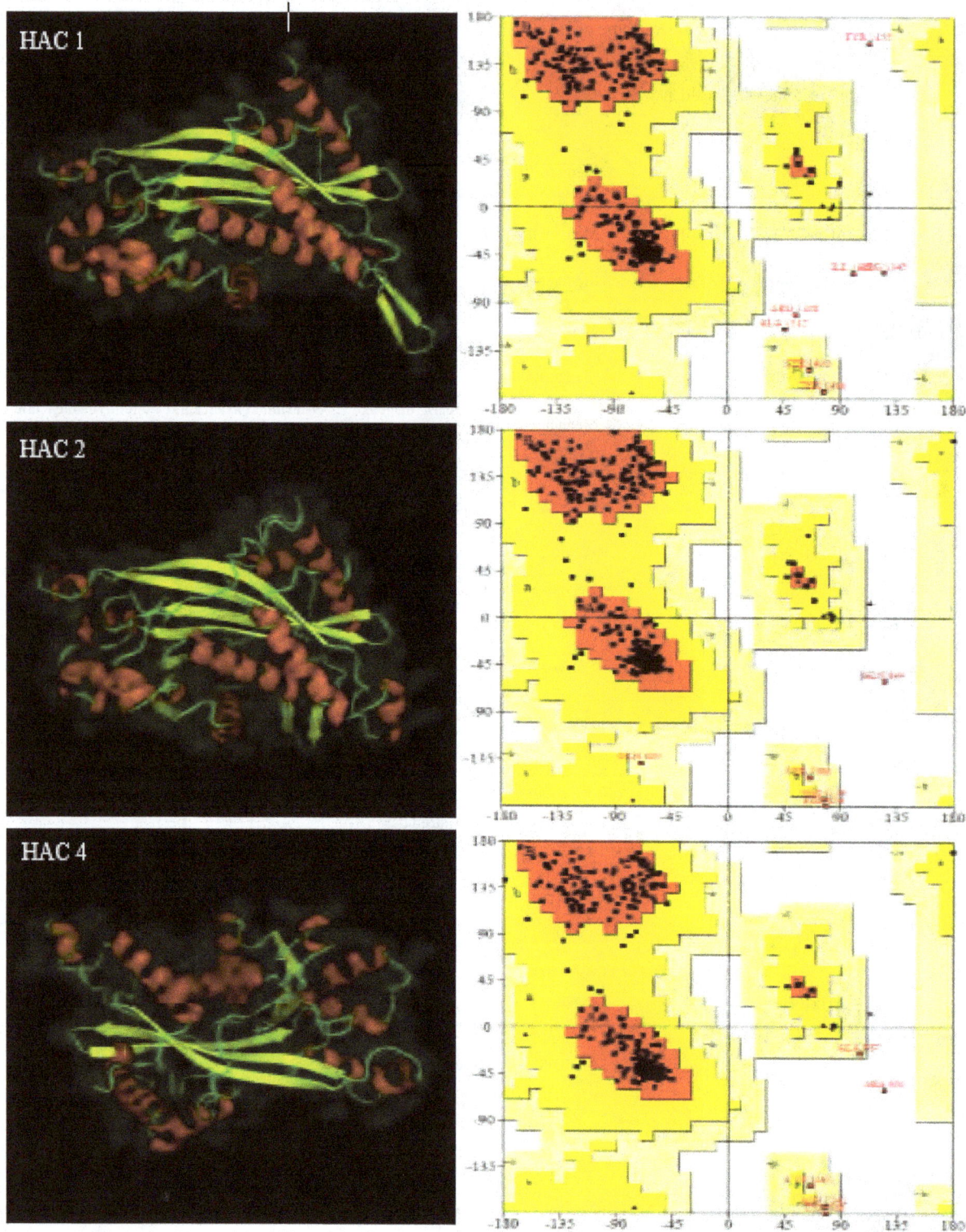

Fig. 1 Predicted 3D structures of HAC1, HAC2 and HAC4 proteins and the respective Ramachandran plots.

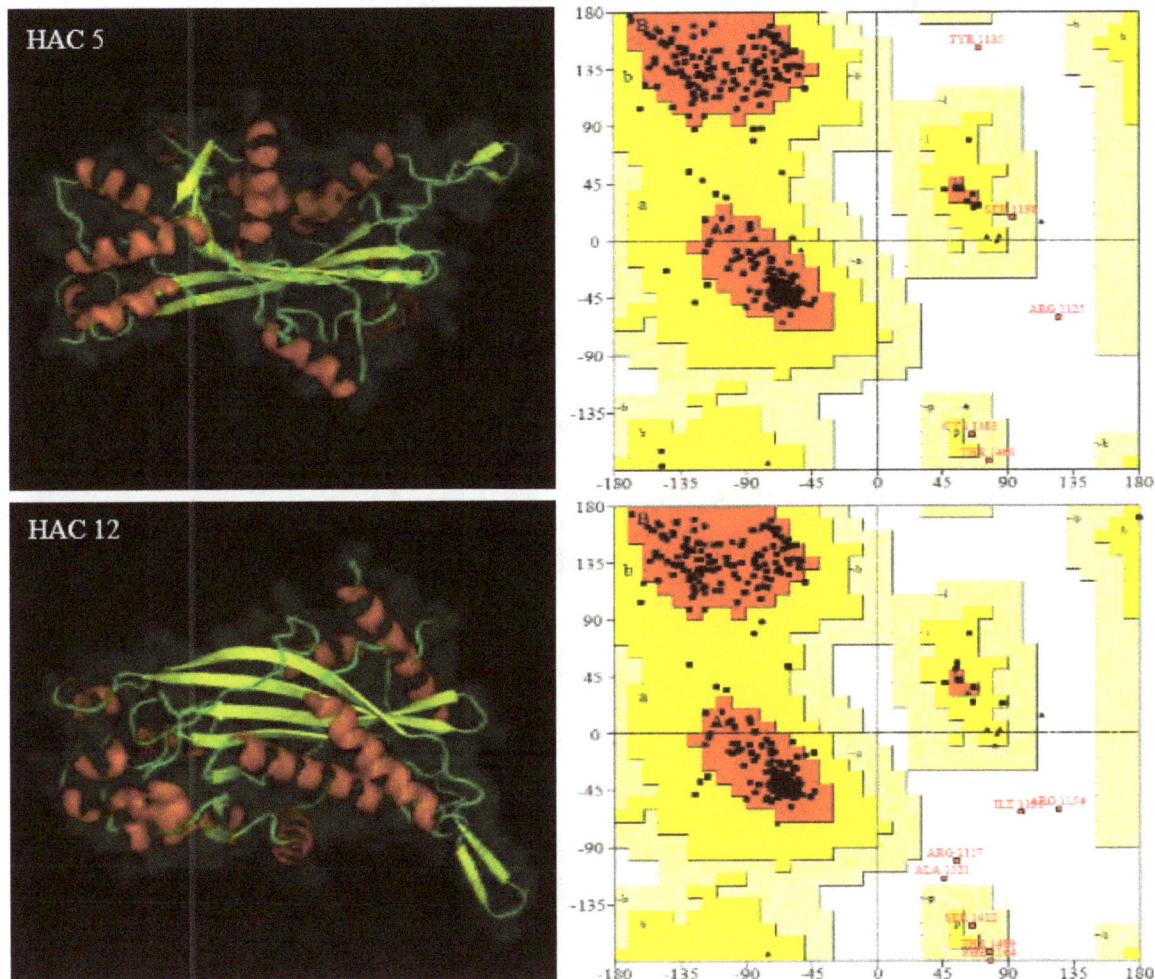

Fig. 2 Predicted 3D structures of HAC5 and HAC12 proteins and the respective Ramachandran plots.

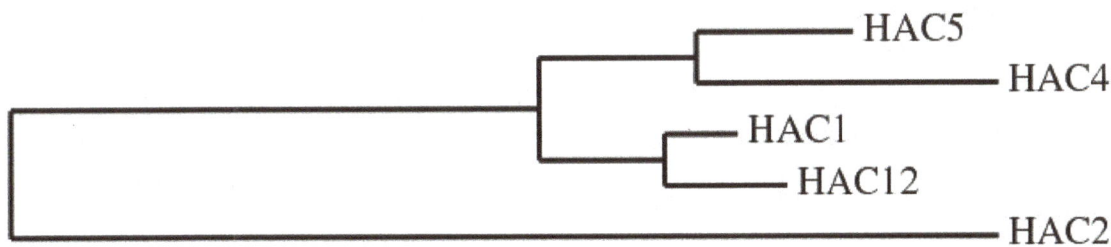

Fig. 3 Phylogenetic tree of the HAC proteins.

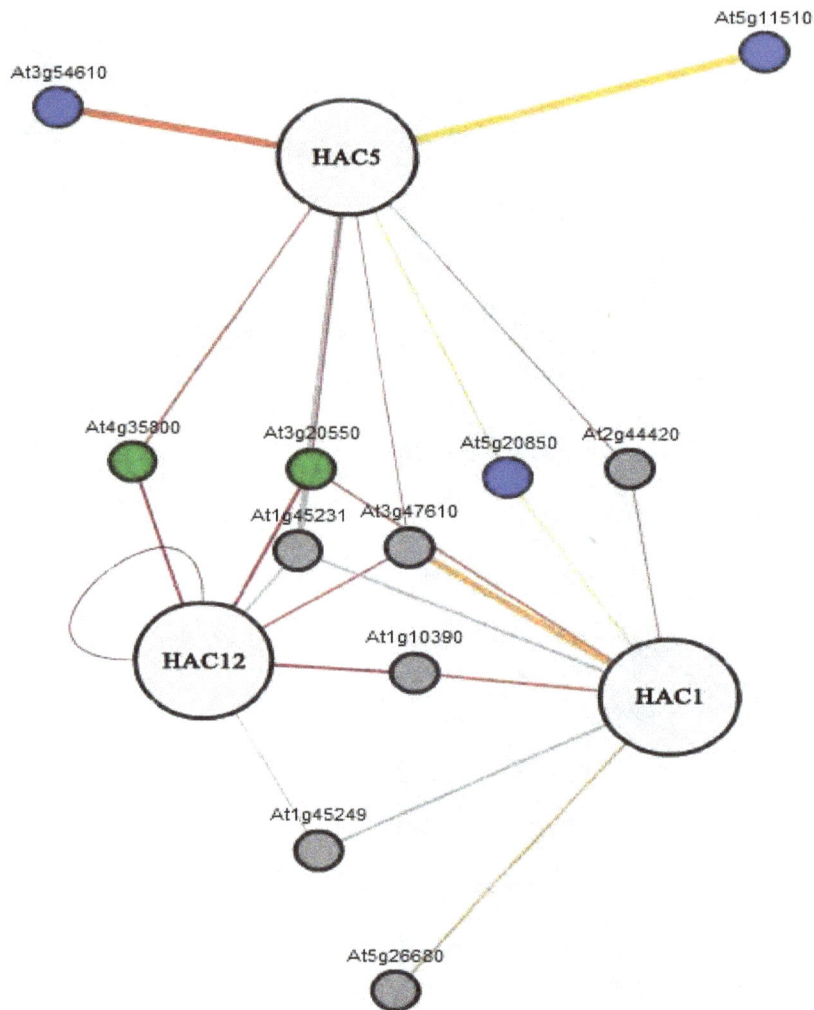

Fig. 4 Interactome of HAC proteins (only significant interaction are shown).

4 Discussion and Conclusion

The five members of the histone CBP acetyltransferase (HAC) protein family in *Arabidopsis thaliana* (HAC1, HAC2, HAC4, HAC5 and HAC12) are plant homologs of the human p300/CBP, involved in transcription regulation, cell-cycle control and differentiation . Its primary function, histone acetylation, is an important posttranslational modification correlated with gene activation. Histone acetylation modifies the protein and lowers the affinity of the respective histone for the DNA molecule, causing the DNA to render and to be accessible to polymerase and transcription factors (Sterner and Berger, 2000). The involvement of p300/CBP and other histone acetyltransferases in processes other than histone acetylation is mostly associated with their interaction with other proteins in the cell. This is important since it allows the assignment of new functions to proteins by analyzing their interactome.

The multiple sequence alignment scores and the phylogenetic tree revealed the evolutionary relationship among HAC homologs. The closest ancestors are HAC1 and HAC12, with an alignment score of 75. This implies that HAC1 and HAC12 have the most recent common ancestor of all HAC proteins. Furthermore, it is noticeable that HAC2 diverged earliest, also reflected in the alignment scores, being HAC2 as the least similar with other HAC homologs. In addition, it can be seen that HAC4 and HAC5 have a relatively high similarity and that they have diverged from HAC1 and HAC12.

By the analysis of the three-dimensional structures of the HAC proteins by aforementioned validation methods, we can conclude that the Phyre2 server created 3D structure models of good quality, confirming the previous in silico structural analysis of Arabidopsis HAC1 (Cemanovic et al., 2014). Phyre2 is based on identifying and aligning remote homologous sequences and relies on profiles or hidden Markov models (HMMs). These profiles/HMMs contain information about the tendency for mutation of each amino acid in a sequence and are unique for each protein. They are created for a set of known 3D structures as well as for the user sequence, and then scanned to find a match (Baum, 1972). Further confirmation and verification of the modeled structure was tested by three validation methods. All the results related to these methods lie in the range of acceptance.

The interactome analysis has linked the HAC1 protein with flap endonuclease-1, which has DNA binding, catalytic and nuclease activity (Tabata, 2000). This may suggest an involvement of HAC1 in these processes, possibly by making the target region of DNA available for flap endonuclease-1 by acetylation of the associated histone proteins.

HAC1 and HAC12 are predicted to interact with nucleoporin autopeptidase, a transporter protein localized in the nuclear membrane (Tamura, 2010). It suggests that this protein transports the newly synthesized proteins into the nucleus where they perform their further functions.

HAC1, HAC5 and HAC12 were all shown to interact with S-adenosylmethionine-dependent methyltransferase domain-containing protein, a methyltransferase enzyme localized mainly in the nucleus. They were also shown to have another common interactome partner - Smad nuclear-interacting protein 1, which functions in RNA and protein binding, and is involved in the processes of DNA methylation, chromatin modification and transcription regulation (Machida and Yuan, 2013). This is in concordance with the findings of Vandel and Trouche (2001), who showed that immunoprecipitation of the human HAC homolog CBP/p300 leads to co-immunoprecipitation of histone methytransferase activity (HMT), and the study of Yang et al. (2009) who proved that CBP/p300 interacts with a methyltransferase enzyme (Protein arginine N-methyltransferase 5). This suggests a link between the processes of acetylation and methylation of histones and an involvement of histone methylation in transcription activation.

HAC1, HAC5 and HAC12 also interact with transcription regulator/ zinc ion binding protein, a transcriptional regulator found in the nucleus. This is in concordance with the fact that the HAC proteins are all involved in the process of transcription regulation. The common interactome protein partners of HAC1, HAC5 and HAC12 can be associated with their high similarity as shown in their alignment score, phylogenetic relationship analysis and their 3D structure similarities.

HAC1 and HAC5 were predicted to interact with N-terminal asparagine amidohydrolase family protein, a protein which also functions in N-terminal protein modification, but by means of myristoylation, i.e. attachment of myristoyl groups to N-termini of proteins (Lin et al., 1999). The two aforementioned HAC proteins also potentially interact with DNA repair protein RAD51-like 1, which is involved in processes of DNA repair, DNA methylation, chromatin modification, transcription- and cell-cycle regulation (Doutriaux et al., 1998). This may also be brought in connection with the findings of Vandel and Trouche (2001) about the potential interaction of acetyltransferases and methyltransferases. Apart from the involvement of RAD51, it is shown that radiation sensitive protein 1 (RAD1) and Replication factor C subunits (RFCs) play an important role in DNA repair and DNA replication processes in *Arabidopsis thaliana* (Abdel Gawwad et al., 2013).

HAC1 and HAC12 have also a possible common interactome partner - abscisic acid-insensitive 5-like protein 5. However, the confidence value for HAC12 is very low in this case, as compared to that of HAC1 (not shown). The interactome partner has DNA binding and protein binding function, and is involved in positive regulation of transcription, similar to the two HAC proteins (Theologis, 2000).

HAC5 and HAC12 were predicted to interact with DNA-directed RNA polymerase II subunit RPB1, the largest subunit of the polymerase II core complex. Besides its involvement in transcription, DNA-directed RNA polymerase II subunit RPB1 is also involved in the process of desumoylation (cleaving of a SUMO – small ubiquitin-related modifier – protein from its target protein) (Nawrath, 1990).

Finally, the strongest confidence values were obtained for the HAC5 interaction with histone acetyltransferase GCN5 and putative c-myb-like transcription factor 3r-4 (MYB3R-4). The former is an acetyltransferase enzyme involved in positive regulation of transcription as well as flower development, like the HAC proteins. In addition, it is also involved in protein ubiquitination and deubiquitination (Servet et al., 2010). This suggests an interaction between HAC5 and GCN5 as DNA modifying enzymes of different protein families at the nuclear level and in the same physiological processes. The latter interactome partner of HAC5, MYB3R-4, is a DNA binding protein involved in transcription regulation, but also in regulation of DNA replication (Haga et al., 2011). A similar interaction is found in humans as well, where the human proto-oncogene c-myb interacts with the CBP/p300 protein (Giordano and Avantaggiati, 1999; Sterner and Berger, 2000). Having a very high confidence value for interacting with MYB3R-4, we can annotate an additional function to HAC5 - regulation of DNA replication.

Conclusively, the analysis of the HAC proteins confirmed their homology and common evolutionary descent. The predicted 3D structures were tested and characterized as being of good quality, and the interactome analysis hinted a possible new functions for some of the HAC proteins.

References

Abdel Gawwad MR, Sutkovic J, Zahirovic E. 2013. 3D structure prediction of replication factor C subunits (RFC) and their interactome in *Arabidopsis thaliana*. Network Biology, 3(2): 74-86

Allfrey VG, Faulkner R, Mirsky AE. 1964. Acetylation and methylation of histones and their possible role in the regulation of RNA synthesis. Proceedings of the National Academy of Sciences USA, 51: 786-794

Anisimova M, Gascuel O. 2006. Approximate likelihood ratio test for branchs: A fast, accurate and powerful alternative. Systems Biology, 55(4): 539-552

Baum LE. 1972. An Inequality and Associated Maximization Technique in Statistical Estimation of Probabilistic Functions of a Markov Process. Inequalities, 3: 1-8

Benkert P, Tosatto SCE, Schomburg D. 2008. QMEAN: A comprehensive scoring function for model quality assessment. Proteins: Structure, Function, and Bioinformatics, 71(1): 261-277

Bernstein FC, Koetzle TF, Williams GJ. 1977. The Protein Data Bank: A Computer-based Archival File for Macromolecular Structures. Journal of Molecular Biology, 112:535

Bordoli L, Netsch M, Lüthi U, 2001. Plant orthologs of p300/CBP: conservation of a core domain in metazoan p300/CBP acetyltransferase-related proteins. Nucleic Acids Research, 29: 589-597

Castresana J. 2000. Selection of conserved blocks from multiple alignments for their use in phylogenetic analysis. Molecular Biology and Evolution, 17(4): 540-552

Cemanovic A, Sutkovic J, Abdel Gawwad MR. 2014. Comparative structural analysis of HAC1 in *Arabidopsis thaliana*. Network Biology, 4(2): 67-73

Chevenet F, Brun C, Banuls A. 2003. TreeDyn: towards dynamic graphics and annotations for analyses of trees. BMC Bioinformatics, 7: 439

Claverie JM, Notredame C. 2007. Bioinformatics for Dummies (Second Edition). Wiley Publishing Inc., Indianapolis, USA

Deng W, Liu C, Pei Y. 2007. Involvement of the Histone Acetyltransferase AtHAC1 inthe Regulation of

Flowering Time via Repression of FLOWERING LOCUS C in *Arabidopsis*. Plant Physiology, 143: 1660-1668

Dereeper A, Audic S, Claverie JM, Blanc G. 2010. BLAST-EXPLORER helps you building datasets for phylogenetic analysis. BMC Evolutionary Biology, 12: 10-18

Dereeper A, Guignon V, Blanc G. 2008. Phylogeny.fr: robust phylogenetic analysis for the non-specialist. Nucleic Acids Research, 36: 465-469

Doutriaux MP, Couteau F, Bergounioux C, White C. 1998. Isolation and characterisation of the RAD51 and DMC1 homologs from *Arabidopsis thaliana*. Molecular Genetics and Genomics, 257: 283-291

Edgar RC. 2004. MUSCLE: multiple sequence alignment with high accuracy and high throughput. Nucleic Acids Research, 32(5): 1792-1797

Eisenberg D, Luthy R, Bowie JU. 1997. VERIFY3D: assessment of protein models with three-dimensional profiles. Methods in Enzymology, 277: 396-404

Geisler-Lee J, O'Toole N, Ammar R. 2007. A Predicted Interactome for *Arabidopsis*. Plant Physiology, 145: 317-329

Giordano A, Avantaggiati ML. 1999. p300 and CBP: partners for life and death. Journal of Cellular Physiology, 181(2): 218-230

Goodman RH, Smolik S. 2000. CBP/p300 in cell growth, transformation, and development. Genes Dev., 14: 1553-1577.

Goujon M, McWilliam H, Li W. 2010. A new bioinformatics analysis tools framework at EMBL-EBI. Nucleic Acids Research, 38: 695-699

Guindon S, Gascuel O. 2003. A simple, fast, and accurate algorithm to estimate large phylogenies by maximum likelihood. Systems Biology, 52(5): 696-704

Haga N, Kobayashi K, Suzuki T. 2011. Mutations in MYB3R1 and MYB3R4 cause pleiotropic developmental defects and preferential down-regulation of multiple G2/M-specific genes in *Arabidopsis*. Plant Physiology, 157(2): 706-17.

Hardin J, Bertoni G, Kleinsmith LJ. 2012. Becker's world of the cell, Eight edition. Benjamin Cummings, San Francisco, USA

Horton P, Park KJ, Obayashi T. 2007. WoLF PSORT: Protein Localization Predictor. Nucleic Acids Research, 1-3 (DOI 10.1093/nar/gkm259)

Hublitz P, Albert M, Peters A. 2009. Mechanisms of Transcriptional Repression by Histone Lysine Methylation. The International Journal of Developmental Biology, 10(1387): 335-354

Katsani KR, Mahmoudi T, Verrijzer CP. 2003. Selective gene regulation by SWI/SNF-related chromatin remodeling factors. Current Topics in Microbiology and Immunology, 274: 113-141

Kelley LA and Sternberg MJE. 2009. Protein structure prediction on the web: a case study using the Phyre server. Nature Protocols, 4: 363-371

Kishimoto M, Fujiki R, Takezawa S. 2006. Nuclear receptor mediated gene regulation through chromatin remodeling and histone modifications. Endocrine Journal, 53(2): 157-172

Lamesch P, Berardini TZ , Li D. 2012. The Arabidopsis Information Resource (TAIR): improved gene annotation and new tools. Nucleic Acids Research, 40: 1202-1210

Larkin MA, Blackshields G, Brown NP. 2002. ClustalW and ClustalX version 2. Bioinformatics, 21: 2947-2948

Laskowski RA, MacArthur MW, Moss DS, Thornton JM. 1993. PROCHECK - a program to check the stereochemical quality of protein structures. Journal of Applied Crystallography, 26: 283-291

Lesk AM. 2002. Introduction to Bioinformatics. Oxford University Press Inc., New York, USA

Letunic I, Doerks T, Bork P. 2012. SMART 7: recent updates to the protein domain annotation resource. Nucleic Acids Research,40: D302-D305

Lin X, Kaul S, Rounsley S. 1999. Sequence and analysis of chromosome 2 of the plant *Arabidopsis thaliana*. Nature, 402 (6763): 761-768

Liu X, Wang L, Zhao K. et al. 2008. The structural basis of protein acetylation by the p300/CBP transcriptional coactivator. Nature, 451: 846-850

Loidl P. 1994. Histone acetylation: facts and questions. Chromosoma, 103(7): 441-449

Machida S, Yuan YA. 2013. Crystal Structure of *Arabidopsis thaliana* Dawdle Forkhead-Associated Domain Reveals a Conserved Phospho-Threonine Recognition Cleft for Dicer-Like 1 Binding. Molecular Plant, [Epub ahead of print].

Martinowich K, Hattori D, Wu H. 2003. DNA methylation-related chromatin remodeling in activity - dependent BDNF gene regulation. Science, 302(5646): 890-893

McGraw S, Vigneault C, Sirard MA. 2007. Temporal expression of factors involved in chromatin remodeling and in gene regulation during early bovine in vitro embryo development. Reproduction, 133(3): 597-608

Nawrath C, Schell J, Koncz C. 1990. Homologous domains of the largest subunit of eucaryotic RNA polymerase II are conserved in plants. Molecular Genetics and Genomics, 223: 65-75

Ogryzko VV, Schiltz RL, Russanova V. 1996. The Transcriptional Coactivators p300 and CBP Are Histone Acetyltransferases. Cell, 87: 953-959

PyMOL. The PyMOL Molecular Graphics System,Version 1.3. Schrödinger, LLC.

Sayers EW, Barrett T, Benson SH. 2009. Database resources of the National Center for Biotechnology Information. Nucleic Acids Research, 37: 5-15

Schultz J, Milpetz F, Bork P, Ponting CP. 1998. SMART, a simple modular architecture research tool: Identification of signaling domains. Proceedings of the National Academy of Sciences USA, 11: 5857-5864

Schwede T, Kopp J, Guex N., Peitsch MC. 2003. SWISS-MODEL: an automated protein homology-modeling server. Nucleic Acids Research, 31: 3381-3385

Servet C, Conde e Silva N, Zhou DX. 2010. Histone acetyltransferase AtGCN5/HAG1 is a versatile regulator of developmental and inducible gene expression in *Arabidopsis*. Molecular Plant, 3(4): 670-677

Sterner DE, Berger SL. 2000. Acetylation of Histones and Transcription-Related Factors. Microbiology and Molecular Biology Reviews, 64(2): 435

Tabata S, Kaneko T, Nakamura Y. 2000. Sequence and analysis of chromosome 5 of the plant *Arabidopsis thaliana*. Nature, 408: 823-826

Tamura K, FukaoY, Iwamoto M. 2010. Identification and Characterization of Nuclear Pore Complex Components in *Arabidopsis thaliana*. Plant Cell, 22: 4084-4097

The UniProt Consortium. 2012. Reorganizing the protein space at the Universal Protein Resource (UniProt). Nucleic Acids Research, 40: D71-D75

Theologis A, Ecker JR, Palm CJ et al. 2000. Sequence and analysis of chromosome 1 of the plant *Arabidopsis thaliana*. Nature, 408(6814): 816-820

Vandel L, Trouche D. 2001. Physical association between the histone acetyl transferase CBP and a histone methyl transferase. EMBO Reports, 2: 21-26

Yang M, Sun J, Sun X et al. 2009. Caenorhabditis elegans Protein Arginine Methyltransferase PRMT-5 Negatively Regulates DNA Damage-Induced Apoptosis. PLoS Genetics, 5(6): e1000514

Yuan LW, Giordano A. 2002. Acetyltransferase machinery conserved in p300/CBP family proteins. Oncogene, 21: 2253-2260

2

Delayed control of ecological and biological networks

Alessandro Ferrarini

Department of Evolutionary and Functional Biology, University of Parma, Via G. Saragat 4, I-43100 Parma, Italy

E-mail: sgtpm@libero.it, alessandro.ferrarini@unipr.it, a.ferrarini1972@libero.it

Abstract

Evolutionary Network Control (ENC) was introduced in 2011 to permit the control of any kind of ecological and biological networks, with an arbitrary number of nodes and links. To date, ENC has been applied with the idea to control biological and ecological networks since the beginning of their system dynamics. This approach has shown to be effective in the control of both continuous-time and discrete-time networks. However a delayed control, where network dynamics are controlled only from a certain point on, could be more economic from a computational viewpoint, and also more feasible from an applicative perspective. For this reason, ENC is further upgraded here to realize the delayed control of ecological and biological nets.

Keywords delayed network control; dynamical networks; genetic algorithms; Euclidean distance; network optimization; system dynamics.

1 Introduction

Evolutionary Network Control (ENC; Ferrarini, 2011) has been conceived to enable the control of any kind of ecological and biological networks, with an arbitrary number of nodes and links, from inside (Ferrarini, 2013) and from outside (Ferrarini, 2013b). The endogenous control requires that the network is optimized at the beginning of its dynamics so that it will inertially go to the desired state. Instead, the exogenous control requires that one or more exogenous controllers act upon the network at each time step.

ENC makes use of an integrated solution (system dynamics - genetic optimization - stochastic simulations) to compute uncertainty about network control (Ferrarini, 2013c) and to compute control success and feasibility (Ferrarini, 2013d). ENC opposes the common idea in the scientific literature that controllability of networks should be based on the identification of the set of driver nodes that can guide the system's dynamics (Ferrarini, 2011), in other words on the choice of a subset of nodes that should be selected to be permanently controlled. ENC has been applied to both discrete-time (systems of difference equations) and continuous-time (systems of differential equations) networks.

ENC employs intermediate control functions to locally (step-by-step) drive ecological and biological networks, so that also intermediate steps (not only the final state) are under its strict control (Ferrarini, 2014).

ENC can also globally subdue nonlinear networks (Ferrarini, 2015), impose early or late stability to any kind of ecological and biological network (Ferrarini 2015b) and locally control nonlinear networks (Ferrarini 2016).

ENC has been also expanded to incorporate the multipurpose control of any kind of ecological and biological network (Ferrarini, 2016b). The rationale is that, not one, but at least two, or even more than two, variables can be contemporaneously driven towards the desired equilibrium values. It is useful whenever ecological and biological networks present several taxonomic resolutions that are worthy to be controlled simultaneously.

A decentralized variant of ENC (Ferrarini, 2016c), where only one node and the correspondent input/output links are controlled, has been also introduced as it could be more economic from a computational viewpoint, in particular when the network is very large (i.e. big data).

A further ENC variant, based on the inhibition of one or several nodes and/or edges, permits to more easily and parsimoniously subdue biological and ecological networks (Ferrarini, 2016d). Another task of ENC is the control of network flows at equilibrium (Ferrarini, 2017).

To date, ENC has been applied with the idea in mind to control biological and ecological networks since the beginning of their system dynamics. However a delayed control, where network dynamics are controlled only from a certain point on, could be more economic from a computational viewpoint, and also more feasible from an applicative perspective. For this reason, ENC has been further upgraded here to realize the delayed control of ecological and biological nets.

Table 1 Evolutionary Network Control (ENC) and its applications.

Reference	Goal
Ferrarini 2011	Theoretical bases of Evolutionary Network Control
Ferrarini 2013	Endogenous control of linear ecological and biological networks
Ferrarini 2013b	Exogenous control of linear ecological and biological networks
Ferrarini 2013c	Computing the uncertainty associated with network control
Ferrarini 2013d	Computing the degree of success and feasibility of network control
Ferrarini 2014	Local control of linear ecological and biological networks
Ferrarini 2015	Global control of nonlinear ecological and biological networks
Ferrarini 2015b	Imposing early/late stability to linear and nonlinear networks
Ferrarini 2016	Local control of nonlinear ecological and biological networks
Ferrarini 2016b	Multipurpose control of ecological and biological networks
Ferrarini 2016c	Decentralized control of ecological and biological networks
Ferrarini 2016d	Structural control of ecological and biological networks
Ferrarini 2017	Control of network flows at equilibrium
This work	Delayed control of ecological and biological networks

2 Delayed Evolutionary Network Control: Mathematical Formulation

An ecological (or biological) dynamical system of n interacting taxonomic resolutions (species, genera, family, etc.) or aggregated assemblages of taxa (e.g., phytoplankton) is as follows

$$\frac{d\mathbf{S}}{dt} = \gamma(\mathbf{S}(t)) \tag{1}$$

where $S_i \, \varepsilon \, \mathbf{S}$ is the number of individuals (or the total biomass) of the generic *i-th* taxonomic resolution (species, genera, family, or aggregated assemblages of taxa). If we also consider inputs (e.g. immigration) and outputs (e.g. emigration) from-to outside, we must write:

$$\frac{d\mathbf{S}}{dt} = \gamma(\mathbf{S}(t)) + \mathbf{I}(t) + \mathbf{O}(t) \tag{2}$$

At the beginning of its dynamics, the network values are

$$\mathbf{S}_0 = <S_1(0), S_2(0)...S_n(0)> \tag{3}.$$

while at the generic time t

$$\mathbf{S}_t = <S_1(t), S_2(t)...S_n(t)> \tag{4}$$

Now let's introduce a desired solution at equilibrium for the dynamics of the studied network

$$\mathbf{S}_d = <S_1(d), S_2(d)...S_n(d)> \tag{5}$$

At each time step, the distance between $\mathbf{S}_t$ and $\mathbf{S}_d$ can be computed as Euclidean distance in the n-dimension space of the n-variable network dynamics:

$$Dist\left(\mathbf{S}_t, \mathbf{S}_d\right) = \sqrt{(S_{1t} - S_{1d})^2 + (S_{2t} - S_{2d})^2 + ... + (S_{nt} - S_{nd})^2} \tag{6}$$

Now, let's introduce a desired threshold distance T_d which triggers the activation of the evolutionary network control (ENC) as follows:

$$\begin{cases} IF & Dist\left(\mathbf{S}_t, \mathbf{S}_d\right) > T_d \\ THEN & \text{ENC is off} \\ ELSE & \text{ENC is on} \end{cases} \tag{7}$$

In other words, ENC is activated if, and only if, the ecological and biological net under study is sufficiently near to the desired solution at equilibrium.

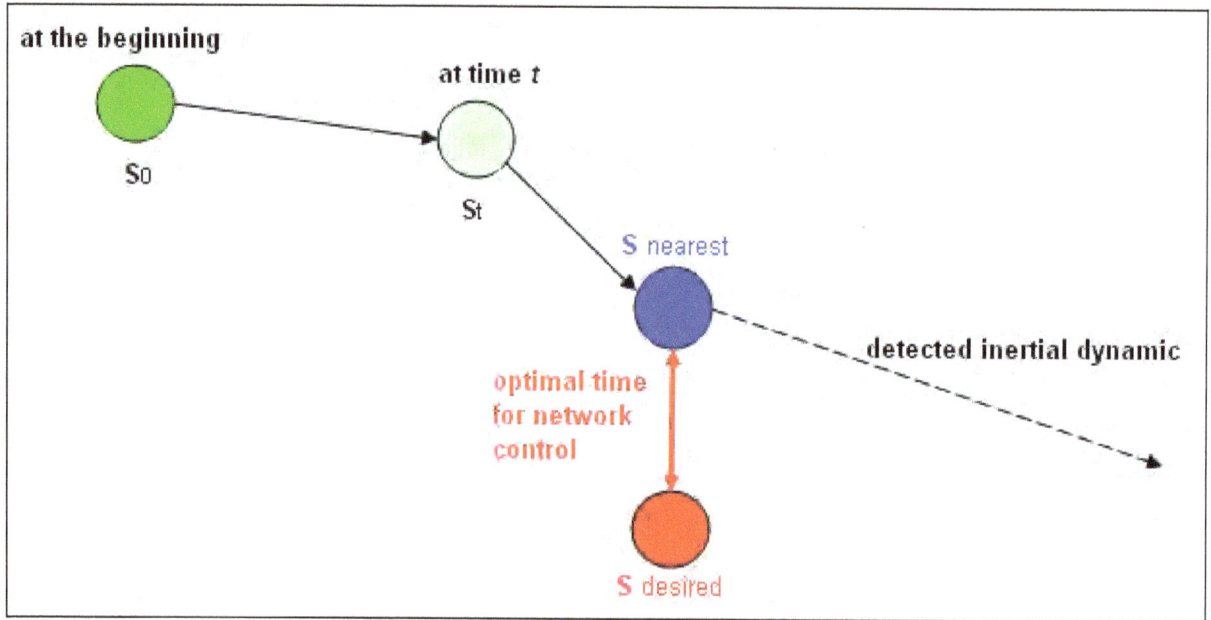

Fig. 1 The idea behind Delayed Evolutionary Network Control (D-ENC). Rather than controlling network dynamics since their beginning, D-ENC seeks the optimal time (T_{opt}) to activate network control. The optimal time happens when network dynamics are closer to the desired solution than a pre-defined threshold value. At that point in time, ENC activates to drive the ecological/biological network towards the desired solution.

Once that ENC is on, it solves the control of Eq. 2 using the following approach (Ferrarini, 2013)

$$\begin{cases} \dfrac{dS_1}{dt} = a_{11*}S_1^* + ... + a_{1n*}S_n^* + I_{1*} + O_{1*} \\ ... \\ \dfrac{dS_n}{dt} = a_{n1*}S_1^* + ... + a_{nn*}S_n^* + I_{n*} + O_{n*} \end{cases} \qquad (8)$$

where any component (variable, parameter or coefficient) of Eq. 8 can be tamed, as denoted by the asterisk, using genetic optimization (Holland, 1975; Goldberg, 1989) to drive the network towards the desired state. ENC can also use an exogenous network control using an external controller C_l (Ferrarini, 2013b)

$$\begin{cases} \dfrac{dS_1}{dt} = a_{11}S_1 + ... + a_{1n}S_n + I_1 + O_1 + c_{11*}C_{1*} \\ ... \\ \dfrac{dS_n}{dt} = a_{n1}S_1 + ... + a_{nn}S_n + I_n + O_n + c_{n1*}C_{1*} \\ \dfrac{dC_1}{dt} = f_1 S_1 + ... + f_n S_n \end{cases} \qquad (9)$$

where asterisks stand for the genetic optimization of exogenous node's edges (i.e., coefficients of interaction with the inner system) and exogenous node's stock. The controller C_l that can also receive feedbacks from the network, which could be subject to control by subduing $<f_1...f_n>$. In case 1 controller is not enough, the model in (9) must be expanded to the following k-external-controllers model (Ferrarini, 2013b):

$$
\begin{cases}
\dfrac{dS_1}{dt} = a_{11}S_1 + ... + a_{1n}S_n + I_1 + O_1 + c_{11*}C_{1*} + ... + c_{1k*}C_{k*} \\
... \\
\dfrac{dS_n}{dt} = a_{n1}S_1 + ... + a_{nn}S_n + I_n + O_n + c_{n1*}C_{1*} + ... + c_{nk*}C_{k*} \\
\dfrac{dC_1}{dt} = f_{11}S_1 + ... + f_{1n}S_n \\
... \\
\dfrac{dC_k}{dt} = f_{k1}S_1 + ... + f_{kn}S_n
\end{cases}
\tag{10}
$$

Many ecological (or biological) dynamical systems can be more properly described using difference (recurrent) equations rather than differential ones. This is true for many systems where dynamics happen on discrete, rather than continuous, time. In this case, the ruling equation becomes

$$
\begin{cases}
(S_1)_{t+1} = a_{11}(S_1)_t + ... + a_{1n}(S_n)_t + (I_1)_t + (O_1)_t \\
... \\
(S_n)_{t+1} = a_{n1}(S_1)_t + ... + a_{nn}(S_n)_t + (I_n)_t + (O_n)_t
\end{cases}
\tag{11}
$$

ENC solves the control of Eq. 11 using the following approach

$$
\begin{cases}
(S_1)_{t+1} = a_{11*}(S_1)_t + ... + a_{1n*}(S_n)_t + (I_1)_{t*} + (O_1)_{t*} \\
... \\
(S_n)_{t+1} = a_{n1*}(S_1)_t + ... + a_{nn*}(S_n)_t + (I_n)_{t*} + (O_n)_{t*}
\end{cases}
\tag{12}
$$

Delayed ENC can be applied using the software Control-Lab 8 (Ferrarini, 2017b).

3 Conclusions

Evolutionary network control (ENC) has been introduced as a methodology where an arbitrary number of network nodes and links can be subdued to drive the network dynamics towards the desired outputs. In previous studies, ENC has shown to be very effective in the control of ecological and biological networks. A delayed control could be more economic from a computational viewpoint, in particular when the network is very large. In this sense, delayed ENC results very promising when applied to big data, the new frontier of network dynamics and control.

References

Ferrarini A. 2011. Some thoughts on the controllability of network systems. Network Biology, 1(3-4): 186-188

Ferrarini A. 2013. Controlling ecological and biological networks via evolutionary modelling. Network Biology, 3(3): 97-105

Ferrarini A. 2013b. Exogenous control of biological and ecological systems through evolutionary modelling. Proceedings of the International Academy of Ecology and Environmental Sciences, 3(3): 257-265

Ferrarini A. 2013c. Computing the uncertainty associated with the control of ecological and biological systems. Computational Ecology and Software, 3(3): 74-80

Ferrarini A. 2013d. Networks control: introducing the degree of success and feasibility. Network Biology, 3(4): 115-120

Ferrarini A. 2014. Local and global control of ecological and biological networks. Network Biology, 4(1): 21-30

Ferrarini A. 2015. Evolutionary network control also holds for nonlinear networks: Ruling the Lotka-Volterra model. Network Biology, 5(1): 34-42

Ferrarini A. 2015b. Imposing early stability to ecological and biological networks through Evolutionary Network Control. Proceedings of the International Academy of Ecology and Environmental Sciences, 5(1): 49-56

Ferrarini A. 2016. Bit by bit control of nonlinear ecological and biological networks using Evolutionary Network Control. Network Biology, 6(2): 47-54

Ferrarini A. 2016b. Multipurpose control of ecological and biological networks. Proceedings of the International Academy of Ecology and Environmental Sciences, 6(3): 75-83

Ferrarini A., 2016c. Decentralized control of ecological and biological networks through Evolutionary Network Control. Network Biology, 6(3): 65-74

Ferrarini A., 2016d. Structural control of ecological and biological networks. Computational Ecology and Software, 6(4): 130-138

Ferrarini A, 2017. A deeper insight into the equilibrium of ecological and biological networks. Network Biology, 7(4): 98-104

Ferrarini A. 2017b. Control-Lab 8: a software for the application of Ecological Network Control. Manual, 108 pages (in Italian)

Goldberg DE. 1989. Genetic Algorithms in Search Optimization and Machine Learning. Addison-Wesley, Reading, USA

Holland JH. 1975. Adaptation in Natural And Artificial Systems: An Introductory Analysis With Applications To Biology, Control and Artificial Intelligence. University of Michigan Press, Ann Arbor, USA

3

Disorder and interactions: What can dehydrins in cereals tell us anymore?

Mouna Choura, Faiçal Brini

Biotechnology and Plant Improvement Laboratory, Centre of Biotechnology of Sfax (CBS)/University of Sfax, B.P "1177" 3018, Sfax Tunisia

E-mail: mouna.choura@cbs.rnrt.tn, faical.brini@cbs.rnrt.tn

Abstract

Dehydrins (DHNs) are intrinsically disordered proteins that are expressed under conditions of water-related stress. They play a fundamental role in plant response and adaptation to abiotic stresses. The protein architecture of dehydrins can be described by the presence of three types of conserved sequence motifs that have been named the Y-, S-and K- segments. Although, dehydrins are extensively studied, their molecular interactions remain elusive. By combining network analysis with prior knowledge, we provide further insights into the role of some dehydrin disorder in cereals notably in stress tolerance. This work includes a comparative analysis with dehydrins of *Arabidopsis thaliana* to highlight the disorder conservation of dehydrins across evolution.

Keywords *Arabidopsis thaliana*; cereals; dehydrins; interactome; intrinsically disordered proteins.

1 Introduction

Dehydrins, or group 2 LEA (Late Embryogenesis Abundant) proteins, constitute a class of intrinsically disordered proteins (IDPs) that are expressed under various water-induced stresses. They accumulate typically in maturing seeds or are induced in vegetative tissues under various stress factors that cause cell dehydration including salinity, dehydration, cold, and freezing stress. Several physiological studies focused on plant stress response have reported a positive relationship between the level of accumulation of dehydrin transcripts or proteins and plant stress tolerance notably for wheat and barley (Kosová et al., 2013). Characteristic of the dehydrins are some highly conserved stretches of seven to 17 residues that are repetitively scattered in their sequences, the Y-, S-, and K- segments. The role of the conserved segments is thus not to promote tertiary structure, but to exert their biological function more locally upon interaction with specific biological targets, for example, by acting as beads on a string for specific recognition, interaction with membranes, or intermolecular scaffolding (Mouillon et al., 2006).

While IDPs such as dehydrins are highly disordered *in vitro*, they often gain structure when bound to a target, suggesting that some disordered proteins may be structured *in vivo* in the presence of their cognate ligands such as membranes. Dehydrin sequences, their structures, and the various ligands that bind to this family of proteins were reviewed by (Graether and Boddington, 2014). The structural physico-chemical and functional characterization of plant dehydrins and how these features could be exploited in improving stress tolerance in plants as outlined by (Hanin et al., 2011). Further works have highlighted the multifunctional roles of dehydrins under environmental stress in plants (Liu et al., 2017). Intrinsically disordered proteins (IDPs) conformational flexibility allows them to recognize and interacts with multiple partners that may increase the interaction speed upon stress (Pietrosemoli et al., 2013).

Currently, applications of protein-protein/gene interaction networks play fundamental role in understanding of the complex biological systems, such as investigation of protein functions and related pathways.

Since the development of the microarray and next generation sequencing technology, more and more transcriptome data has become available. The high throughput computational methods have been successfully used to predict some interactomes such as *A. thaliana* (Consortium, 2011), *O. Sativa* (Ho et al., 2012), *Z. mays* (Zhu et al., 2016) and *S. Bicolor* (Tian et al., 2016). Due to the information provided by related databases and public papers, it is a benefit to investigate the available dehydrin interaction networks to get further insights into their role in stress tolerance.

2 Material and Methods

2.1 Protein disorder prediction

The prediction of intrinsic protein disorder was carried out using the reputable and accurate IUPred web server (http://iupred.enzim.hu/) (Dosztányi et al., 2005). This server disordered region from amino acid sequence based on pairwise energy content. There are three different prediction types, each using different parameters optimized for slightly different applications. These are: long disorder, short disorder, and structured domains.

2.2 Protein binding region prediction

The binding regions involved in protein-protein interaction were predicted using ANCHOR tool (http://anchor.enzim.hu/) (Dosztányi et al., 2009), which is based on the IUPred program mentioned above. Anchor predicts binding regions located in disordered proteins from the amino acid sequence.

2.3 Disorder evaluation in protein-protein interaction

We extracted data on binary protein-protein interactions in *O.sativa*, *Z.Mays*, *S.bicolor* and *A.thaliana* from the STRING database (Szklarczyk et al., 2015).

3 Results and Discussion

3.1 The structural characteristics of dehydrins: disorder and function

Dehydrins share a number of structural features. One of the most notable features is the presence, in their central region, of a continuous run of five to nine serines followed by a cluster of charged residues. Such a region has been found in all known dehydrins so far with the exception of pea dehydrins. A second conserved feature is the presence of two copies of K-segments. The first copy is located just after the cluster of charged residues that follows the poly-serine region and the second copy is found at the C-terminal extremity. The presence of the K segment raises the question of whether DHNs bind lipids, bilayers, or phospholipid vesicles. DHN1 in *Zea mays* (*ZmDHN1 or RAB17*) can bind to lipid vesicles that contain acidic phospholipids. It was reported that *ZmDHN1* binds more favourably to vesicles of smaller diameter than to larger vesicles, and that the association of DHN1 with vesicles results in an apparent increase of α-helicity of the protein.

Thus, the role of the K-segment may be hydropobic interaction with partially denatured proteins and protect the cell membranes, especially under the stressed conditions (Koag et al., 2003, 2009).

Rice (*O. sativa* L.) is susceptible to drought-induced stress as compared to other cereals (Mostajeran AND Rahimi-Eichi, 2009), resulting in poor seedling vigor (Bouman et al., 2005), fertility and adversely affecting the crop yield. It has been reported that *OsDHN1* gene overexpressed in rice confers high tolerance to drought and salt stress (Kumar et al., 2014). Further, rice dehydrin K-segments are responsible for the antibacterial activities against Gram positive bacteria (Zhai et al., 2011). Previous data showed that YSK2 type dehydrin from *Sorghum bicolor* (*SbDhn1*) play a protective role under high temperature and osmotic stress. *SbDhn2* possessed metal binding as well as radical scavenging activity (Halder et al., 2017) (Halder et al., 2018). Another study revealed that the K-segments of the wheat dehydrin DHN-5 are essential for the protection of enzyme activities *in vitro*. In addition, the K-segments have antibacterial and antifungal activities against Gram-positive and Gram-negative bacteria and fungi (Drira et al., 2015).

3.2 Dehydrins: disorder and interactions

The analysis of the amino composition of the dehydrin family (Pfam00257) reveals that dehydrins (DHN1) in *O. sativa*, *Z. mays*, *S. bicolor*, *T. aestivum*, *T. turgidium* and *A. thaliana* have similar disorder percentage (Fig. 1). Moreover, similar disordered binding regions (DBR) were noted (Table 1). Interestingly, these regions are involved in DNA, metal ions or lipid vesicles bindings under stress (Liu et al., 2017). This may explain that no physical partner was found in the network. In addition, the *A. thaliana* interactome is more studied than those of cereals.

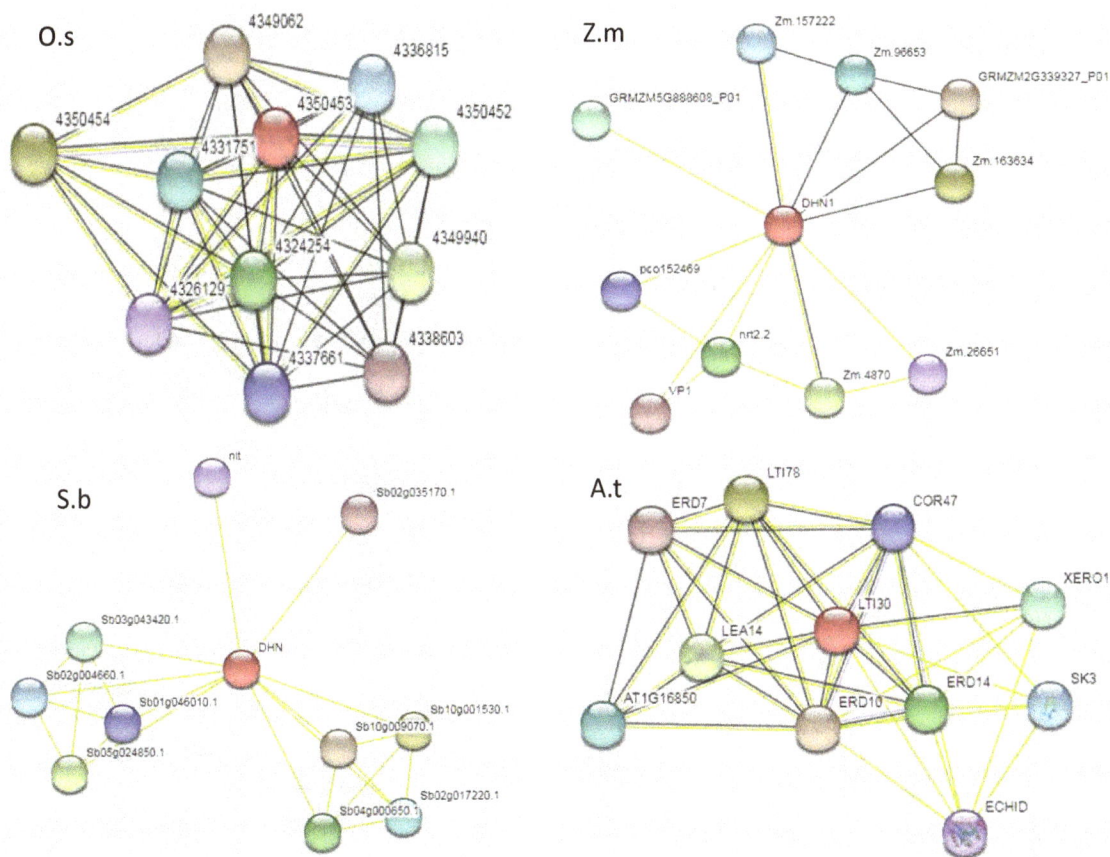

Fig. 1 Protein-Protein interaction networks of *DHN1* in *O. sativa*, *Z. mays*, *S. bicolor* and *A. thaliana* generated by STRING software (http://string-db.org/). Functional partners are detailed in Table 1.

Table 1 List of some dehydrins, disorder percentages, the number of disorder binding regions and the predicted functional partners.

Organism	name	Accession	% of Disorder	DBR	Nb of partners	Predicted functional partners
A. thaliana	Xero2	P42758	33	4	10	-ERD10: *Dehydrin ERD10* -LTI78: LOW-TEMPERATURE-INDUCED 78 -LEA14: LATE EMBRYOGENESIS ABUNDANT 14 -ERD14: Dehydrin ERD14 -Dehydrin XERO1 - AT1G16850 : uncharacterized protein -SK3 : *SKP1-like 3* - COR47 :Dehydrin COR47 -ECHID: *Naphthoate synthase* -ERD7:EARLY-RESPONSIVE TO DEHYDRATION 7
O. sativa	Rab16B	Q2R4Z5	83	4	10	- 4349062 :Expressed protein - 4350454 :Dehydrin putative - 4349940 :expressed protein - 4324254 : Late embryogenesis abundant protein, group 3, putative, expressed - 4350452: Dehydrin putative expressed - 4331751: Embryonic protein DC-8, putative, expressed - 4336815 : Late embryogenesis abundant group 1, putative, expressed - 4337661 : Putative low temperature and salt responsive protein, expressed - 4326129: Late embryogenesis abundant protein, group 3, putative, expressed - 4338603 : Eukaryotic peptide chain release factor subunit 1-1, putative, expressed
T.aestivum	LEA2	U6C7L2	85	3	-	
T.turgidum	Dehydrin	Q6IVU6	74.01	4	-	
Z.mays	ZmDHN1 (RAB17)	P12950	92	4	10	-GRMZM2G339327_P01: Putative uncharacterized protein -Zm.163634: Putative uncharacterized protein - Zm.4870 : dehydrin - nrt2.2 : Putative high affinity nitrate transporter - GRMZM5G888608_P01: Putative uncharacterized protein - Zm.96653 : Putative uncharacterized protein - Zm.157222: Putative uncharacterized protein - pco152469 :stress protein - Zm.26651 : Late embryogenesis abundant protein -VP1: Regulatory protein viviparous-1
S.bicolor	SbDHN	C5YX70	100	4	10	- Sb10g009070 : Putative uncharacterized protein - Sb10g001530 : Putative uncharacterized protein - Sb04g000650.1 : Putative uncharacterized protein - Sb03g043420.1 : Putative uncharacterized protein - Sb02g017220.1 : Putative uncharacterized protein - Sb02g004660.1: putative glycosyl-hydrolase - Sb01g046010 :putative sugar transport protein - nit: nitralse - Sb02g035170.1 : putative translation initiation factor SUI1

3.3 Gene interaction network

Currently, applications of protein-protein/gene interaction networks play fundamental role in understanding of the complex biological systems, such as in investigating of protein functions and related pathways. Intrinsically disordered proteins (IDPs) conformational flexibility allows them to recognize and interacts with multiple partners, that may increase the interaction speed (Pietrosemoli et al., 2013). Based on genomic context, high-throughput experiments, co-expression and text mining, gene co-expression networks of *OsDHN1, ZmDHN1, SbDHN1, LEA2 and Xero2* were constructed (Fig. 2). In the network, nodes represent genes and edges represent the interaction. The degree of a node is the number of edges connected to this node. The clustering coefficient of a node is the ratio of the observed number of direct connections between the node's immediate network neighbours over the maximum possible number of such connections.

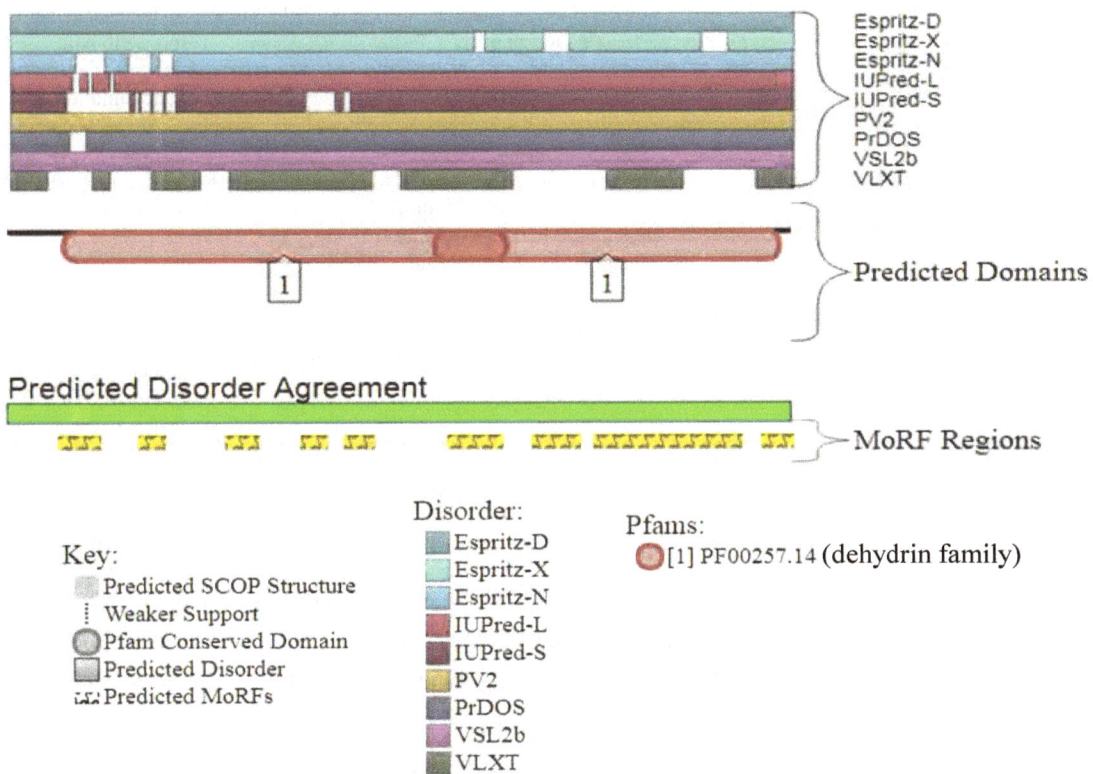

Fig. 2 Intrinsic disorder propensity and some important disorder-related functional information generated for dehydrin in wheat by the D2P2 database (http://d2p2.pro/) (Oates et al., 2013). Here, the green bar in the middle of the plot shows the predicted disorder agreement between nine predictors. Yellow bars show the location of the predicted disorder-based binding sites (molecular recognition features, MoRFs).

The comparison of networks indicated similar topological features (Table 2). Moreover, functional features of networks revealed that dehydrins are mainly involved in water stress (GO: 0009415) and in response to stress (GO: 0006950) (Table 3). This functional conservation should guide further effort to explore the role of dehydrin in cereals. Although the molecular function of the dehydrins still unknown, dehydrins bind to various molecules such as water, phospholipids in maize and in metals in *A. thaliana*. These features explain the proposed functions of dehydrins notably protective activity, membrane stability and metal binding involving interaction to the cellular targets often without partners.

Table 2 Network statistics.

Dehydrin	Nb of nodes	Nb of edges	Avg.node degree	Avg. local clustering coefficient
OsDHN1	11	49	8.91	0.904
ZmDHN1	11	18	3.27	0.819
SbDHN	11	22	4	0.933
Xero2	11	39	7.09	0.847

Table 3 The gene ontology (GO) molecular functions and biological processes of dehydrins in *A.thaliana, O.sativa, Z. mays, S.bicolor, T.aestivum, H.vulgare.*

Organism	Molecular Function	Biological process
A.thaliana	cobalt ion binding	cold acclimation
	copper ion binding	defense response to fungus
	nickel cation binding	response to abscisic acid
	zinc ion binding	response to cold
		response to water deprivation
O.sativa	unknown	cold acclimation
		response to abscisic acid
		response to water deprivation
Z. mays	phosphatidic acid binding	Response to stress
	phosphatidylglycerol binding	Response to water
	phosphatidylinositol binding	
	phosphatidylserine binding	
S.bicolor	unknown	Response to stress
		Response to water
T.aestivum	unknown	Response to stress
		Response to water
H.vulgare	unknown	Response to stress
		Response to water

4 Conclusion

This study has focused on some dehydrins in cereals and *A. thaliana* in order to investigate the potential roles of disorder in the function and molecular interactions. Leveraging to their high structure flexibility, they stabilize membrane and cellular targets under extreme environmental conditions such as low temperatures and drought. Although the data limitations of some cereal interactomes such as wheat and barley, the network based approach provides a global view of relationships between dehydrins and their potential partners. Function similarities between dehydrins in cereals and *A. thaliana* may suggest similar partners. The large data observed to date predicts that many novel dehydrin partners remains to be identified and validated.

Acknowledgements

This study was supported by the Ministry of Higher Education and Scientific Research of Tunisia.

References

Bouman BAM, Peng S, Castaneda AR, Visperas RM. 2005. Yield and water use of irrigated tropical aerobic rice, systems. Agricutural Water Management, 74(2): 87-105

Consortium AIM. 2011. Evidence for network evolution in an Arabidopsis interactome map. Science, 333 (6042): 601-607

Dosztányi Z, Csizmok V, Tompa P, Simon I. 2005. IUPred: web server for the prediction of intrinsically unstructured regions of proteins based on estimated energy content. Bioinformatics, 21(16): 3433-3434

Dosztányi Z, Mészáros B, Simon I. 2009. ANCHOR: web server for predicting protein binding regions in disordered proteins. Bioinformatics, 25(20): 2745-2746

Drira M, Saibi W, Amara I, Masmoudi K, Hanin M, Brini F. 2015. Wheat Dehydrin K-Segments Ensure Bacterial Stress Tolerance, Antiaggregation and Antimicrobial Effects. Applied Biochemistry and Biotechnology, 175(7): 3310-3321

Graether SP, Boddington KF. 2014. Disorder and function: a review of the dehydrin protein family. Frontiers in Plant Science, 5: 576

Halder T, Upadhyaya G, Ray S. 2017. YSK$_2$ type dehydrin (*SbDhn1*) from *Sorghum bicolor* showed improved protection under high temperature and osmotic stress condition. Fronters in Plant Science, 8: 918

Halder T, Upadhyaya G, Basak C, Das A, Chakraborty C, Ray S. 2018. Dehydrins Impart Protection against Oxidative Stress in Transgenic Tobacco Plants. Frontiers in Plant Science, 9: 136

Hanin M, Brini F, Ebel C, Toda Y, Takeda S, Masmoudi K .2011. Plant dehydrins and stress tolerance: versatile proteins for complex mechanisms. Plant Signaling and Behavior, 6(10): 1503-1509

Ho CL, Wu Y, Shen HB, Provart N, Geisler M. 2003. A predicted protein interactome for rice. Rice, 5(1): 15

Koag MC, Fenton RD, Wilkens S, Close TJ .2003. The binding of maize DHN1 to lipid vesicles. Gain of structure and lipid specificity. Plant Physiology, 131(1): 309-3016

Koag MC, Wilkens S, Fenton RD, Resnik J, Vo E, Close TJ .2009. The K-segment of maize DHN1 mediates binding to anionic phospholipid vesicles and concomitant structural changes. Plant Physiology, 150(3): 1503-1514

Kosová K, Vítámvás P, Prášilová P, Prášil IT .2013. Accumulation of WCS120 and DHN5 proteins in differently frost-tolerant wheat and barley cultivars grown under a broad temperature scale. Biologia Plantarum, 57(1): 105-112

Kumar M, Lee SC, Kim JY, Kim SJ, Aye SS, Kim SR. 2014. Over-expression of dehydrin gene, OsDhn1, improves drought and salt stress tolerance through scavenging of reactive oxygen species in rice (Oryza sativa L.). Journal of Plant Biology, 57(6): 383-393

Liu Y, Song Q, Li D, Yang X, Li D .2017. Multifunctional roles of plant dehydrins in response to environmental stresses. Frontiers in Plant Science, 8: 1018

Mostajeran A, Rahimi-Eichi V. 2009. Effects of drought stress on growth and yield of rice (*Oryza sativa L.*) cultivars and accumulation of proline and soluble sugars in sheath and blades of their different ages leaves. American-Eurasian Journal of Agriculture and Environmental Sciences, 5(2): 264-272

Mouillon JM, Gustafsson P, Harryson P. 2006. Structural investigation of disordered stress proteins. Comparison of full-length dehydrins with isolated peptides of their conserved segments. Plant Physiology, 141(2): 638-650

Pietrosemoli N, García-Martín JA, Solano R, Pazos F. 2013. Genome-wide analysis of protein disorder in Arabidopsis thaliana: implications for plant environmental adaptation. Plos One, 8(2): e55524

Szklarczyk D, Franceschini A, Wyder S, Forslund K, et al. 2015. STRING v10: protein-protein interaction networks, integrated over the tree of life. Nucleic Acids Res, 43(database issue): D447-452

Tian T, You Q, Zhang L, Yi X, et al. 2016. Sorghum FDB: sorghum functional genomics database with multidimensional network analysis. Database, baw099 (doi: 10.1093/database/baw099)

Zhai C, Lan J, Wang H, Li L, Cheng X, Liu G. 2011. Rice dehydrin K-segments have *in vitro* antibacterial activity. Biochemistry, 76(6): 645-650

Zhu G, Wu A, Xu XJ, Xiao PP, Lu L, Liu J, Cao Y, Chen L, Wu J, Zhao XM . 2016. PPIM: A Protein-Protein Interaction Database for Maize. Plant Physiology, 170(2): 618-626

4

Drug design and analysis for bipolar disorder and associated diseases: A bioinformatics approach

Nahida Habib[1], **Kawsar Ahmed**[2,3], **Iffat Jabin**[1], **Mohammad Motiur Rahman**[1]

[1]Department of Computer Science and Engineering (CSE), Mawlana Bhashani Science and Technology University (MBSTU), Santosh, Tangail-1902, Bangladesh

[2]Department of Information and Communication Technology (ICT), Mawlana Bhashani Science and Technology University (MBSTU), Santosh, Tangail-1902, Bangladesh

[3]Group of Bio-photomatix, Santosh, Tangail-1902, Bangladesh

E-mail:nahidahabib164@yahoo.com, kawsar.ict@mbstu.ac.bd, k.ahmed.bd@ieee.org, kawsarit08050@gmail.com, iffat.jabin02 @gmail.com, mm73rahman@gmail.com

Abstract

Bioinformatics deals with biological data and analyzes or processes the data using computer science techniques. With the appearance of modern bioinformatics tools, it is now possible to design a drug using these high technologies and open a new area of drug design and development. This research predicts to design a common drug for four associated mental disorders that include bipolar disorder, schizophrenia, coronary heart diseases and stroke. The key to drug design is a biomolecule or protein. To show the protein interactions and evolutions, a protein–protein interaction network is created among the common genes of the four diseases. The genes corresponding to each disease are collected from NCBI gene database. These genes are preprocessed, mined and verified to find the common genes among the diseases. After getting common genes (7 genes), PPI network is created with them. Then a common drug is designed that will work on four investigated diseases. This structure based drug design research will open a new era to discover and develop new drug compounds using different bioinformatics tools.

Keywords bioinformatics; mental disorders; bipolar disorder; schizophrenia; coronary heart diseases; stroke; PPI Network; drug design.

1 Introduction

Bioinformatics plays an important role in data processing, sequencing, handling large dataset, storing, transformation, visualizing PPI network and last but not least drug design. Modern bioinformatics tool made it easier to design a common drug for bipolar disorder and associated diseases. Drug targets are typically key molecules involved in a specific metabolic or cell signaling pathway that is known, or believed, to be related to a particular disease state and are most often proteins and enzymes in these pathways (Zhang, 2016b; TechTarget, 2017; Zhang and Feng, 2017). A common pathway shared (Nahida, 2016), by bipolar disorder, schizophrenia, coronary heart diseases and stroke can be designed to show the association and interaction among the proteins of these diseases which leads to the way to drug design. Moreover, designing a common drug for four associated diseases will decrease the amount of drug one should absorb for the diseases separately.

Mental disorder, mental illness, major depressive disorder or psychiatric disorder include a wide range of problems, such as depression, major depression, bipolar disorder, schizophrenia, coronary heart diseases, stroke, anxiety disorders etc. There may be multiple causes of mental disorders. Genes, biological factors, environment and lifestyle, family history, a stressful job or home life, traumatic brain injury, mother's exposure to viruses or toxic chemicals while pregnant may play a part (NIH). One in 5 adults experiences a mental health condition every year and one in 17 lives with a serious mental illness such as schizophrenia or bipolar disorder (Nami National Alliance on Mental Illness). Childhood disorders can lead to adult disorders or personality disorders.

Bipolar disorder and schizophrenia are among the most severe and complex mental disorders. Bipolar disorder-I (BD-I) is a highly prevalent and often chronic mood disorder with a lifetime prevalence of 1% to 5% (Akiskal et al., 2000), which is frequently characterized by episodic recurrent mania or hypomania and major depression (Belmaker, 2004). Bipolar disorder (BD) that is known as manic-depressive illness, affects approximately 1% of the general population (Merikangas et al., 2011) with high heritability (Edvardsen et al., 2008). Schizophrenia is a devastating disorder affecting ~1% of the population (Ayalew, 2012). Overall, the incidence of schizophrenia was found to be higher in males than females (McGrath et al., 2004). The ratio of incidence rates between men and women was 1.4 (Aleman et al., 2003).

CHD and stroke are also complex disorders. Coronary heart disease (CHD) is a leading cause of death in the United States (Keenan, 2011) and the world (Lozano et al., 2012). Within the studies focusing on medical illnesses among patients with bipolar disorder, the most common medical problems cited are obesity, diabetes mellitus and subsequent cardiovascular disease; all of these medical conditions, as well as depressive symptoms, are recognized as risk factors for stroke (Everson et al., 1998; Krishnan, 2005). Nevertheless, there is scant information on the risk of developing stroke among patients with bipolar disorder, despite cerebrovascular diseases having been reported as one of the major causes of death among this particular patient population (Hoyer et al., 2000; Joukamaa et al., 2001; Tsai et al., 2005). CVD is the leading cause of death in BD, with a standardized mortality ratio of 1.5 to 2.5 (Osby, 2001; Weeke, 1987). Approximately 38% of all individuals who suffer a heart attack will die from it (Rosamond et al., 2008).

This research investigates with BD, SZ, CHD and stroke whose are directly or indirectly associated. For the first time we tried to design a common drug for these associated diseases in this research. This paper is organized in 5 sections. Section 2 discusses about the background or previous works related to the research, section 3 describes the proposed methodology and working principle, section 4 discusses and analyzes the result and last but not least section 5 includes conclusion and future work.

2 Background

Prior research shows the interaction between the investigated four diseases (BD, SZ, CHD and stroke). Also different bioinformatics tools were used for this purpose. Information about the bio-informatics tools that can be used to show interaction, finding common genes and represent the PPI network among genes as well as proteins are described in the article (Klingstrom and Plwczynski, 2010). It has recently been shown that mental diseases like schizophrenia, major depression and bipolar disorder may be quite closely related genetically, meaning that there is a considerable genetic overlap between the three different diagnoses (Rasic, 2013). Based on this recent meta-analysis it was found that offspring of parents with one of the mentioned diseases not only had an increased risk of developing the same illness as their ill parent, but also an increased risk of developing the other two disorders (Thorup et al., 2015). The detail information about associations of schizophrenia and psychotic bipolar disorder for default mode network were provided on the article by Medaa (2014).

There is substantial evidence for partial overlap of genetic influences on schizophrenia and bipolar disorder, with family, twin, and adoption studies showing a genetic correlation between the disorders of around 0.6 (Cardno and Owen, 2014). The article (Osby et al., 2001), said that patients with BD are up to twice as likely to die from cardiovascular causes as their counterparts in the general population. The association between major depression, bipolar disorder & cardiovascular diseases are shown in the paper (Baune et al., 2006). The article (Sowden and Huffman, 2009) interpreted that depression, anxiety disorders, schizophrenia and bipolar disorder (BD) have all been identified as risk factors for the onset and progression of cardiovascular disease (CVD). Individuals with chronic heart diseases or stroke have a significantly increased incidence and prevalence of affective disorders (Baune et al., 2006). The paper (De Hert et al., 2009), mentioned that compared with healthy controls, people with SMI who were not prescribed any antipsychotics were at increased risk of CHD and stroke than controls, whereas those prescribed such agents were at even greater risk. The paper also revealed that those receiving the higher doses were at greatest risk of death from both CHD and stroke. In the United States alone, more than 16.3 million adults have CHD and an estimated 935,000 heart attacks occur each year (Roger et al., 2012). With regard to vascular outcomes, lifetime rates of stroke, myocardial infarction, and other forms of CVD are elevated among people who report having been maltreated as children (Goldstein et al., 2015).

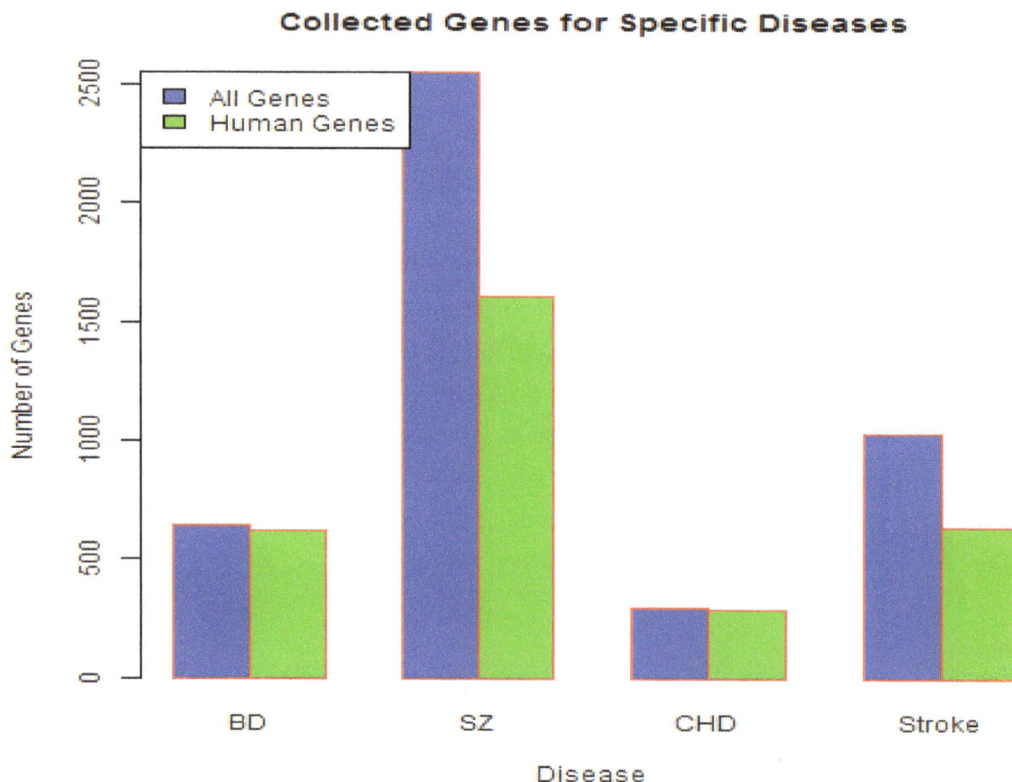

Fig. 1 Collected genes for specific diseases.

Jesmin et al. (2016) provided a common disease regulatory network for metabolic disorders by investigating on associated diseases. The UNIHI tool that is used to predict PPI network, common metabolic pathway and drug design is referred by Kalathur et al. (2014). The role of gene duplication for creating gene network evolution has been investigated in the paper of Teichmann and Babu (2014). In addition, Li and Zhang (2013) identified crucial metabolites/reactions in tumor signaling networks. Zhang (2016a) developed a mathematical model to describe dynamics of occurrence probability of missing links in predicted missing link list. Zhang and Feng (2017) used network analysis to analyze metabolic pathway of non-alcoholic fatty liver disease.

Various previous research have been held on bioinformatics that includes analyzing genes, finding PPI network and common pathway for associated diseases. The presented research is a descendant of previous research which aimed to design a common drug for BD, SZ, CHD and stroke.

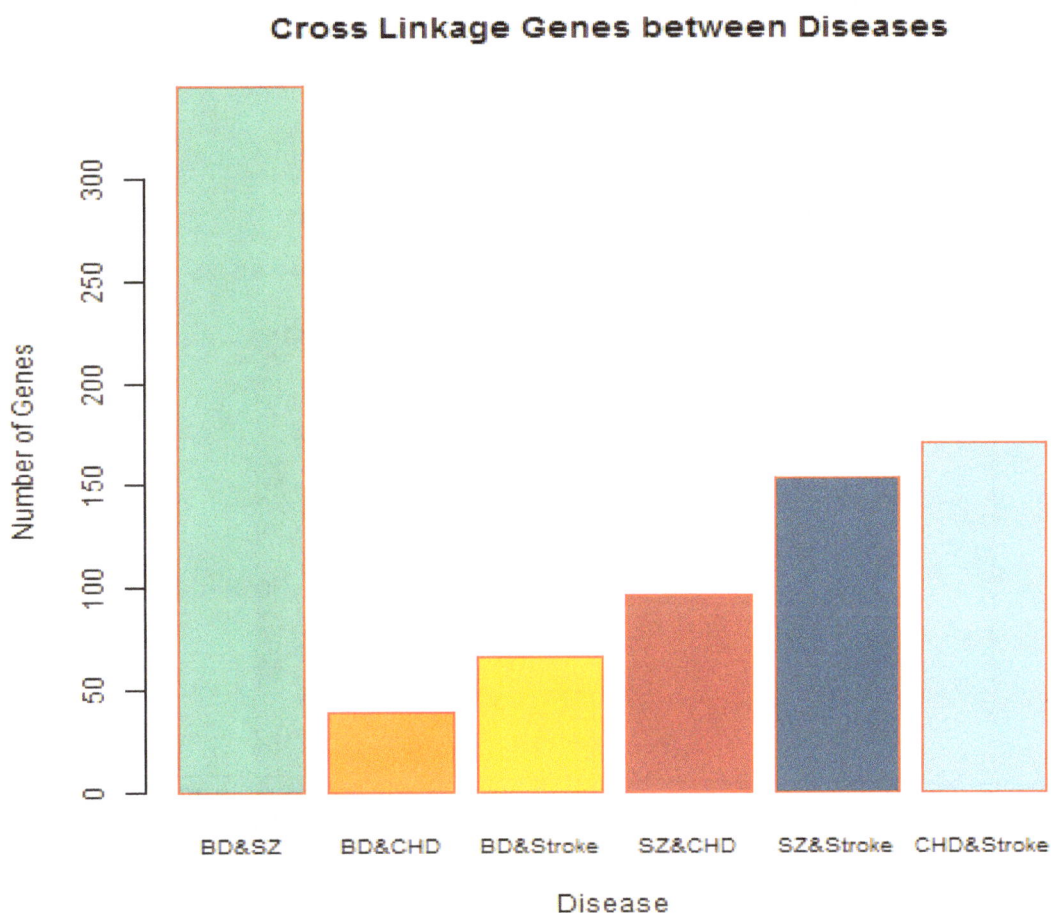

Fig. 2 Cross linkage genes between selected diseases.

3 Proposed Methodology

Drug design is a step by step process. Several steps are performed in this section to reach the desire goal. Each of the steps is described below in the following subsections through 3.1 to 3.4 respectively.

3.1 Gene filtering

The NCBI (National Center for Biotechnology Information) is an important resource for bioinformatics tools and services. It maintains a huge database of all the DNA and protein sequence data. For this research project genes associated with bipolar disorder, schizophrenia, coronary heart disease and stroke diseases are collected from NCBI gene database (Nahida, 2016). Collected genes are preprocessed and filtered and the genes only responsible for *Homo sapiens* are stored for further processing.

3.2 Gene mining

Gene mining is one of the most important parts of this research. Corresponding genes are downloaded in the sorted increasing order by their weight. The collected genes for each disease are merged to find gene linkage. The interrelated genes between 2 selected diseases (like BD & CHD; BD & Stroke etc.) and among 3 selected diseases (like BD, CHD & stroke; BD, CHD & schizophrenia, etc.) are identified and collected. From the sorted linkage gene files only the genes are mined (Nahida, 2016). Then the common genes from top 100 and top 50 interrelated genes are searched and collected.

3.3 PPI network

Protein-Protein Interaction Network or PPI network is used to show the protein interaction and common pathway among the interrelated genes. UniHi, a very popular reliable bioinformatics tools is used for this purpose. In this step, from the interrelated common genes, PPI networks and common pathways are created using UniHi tool.

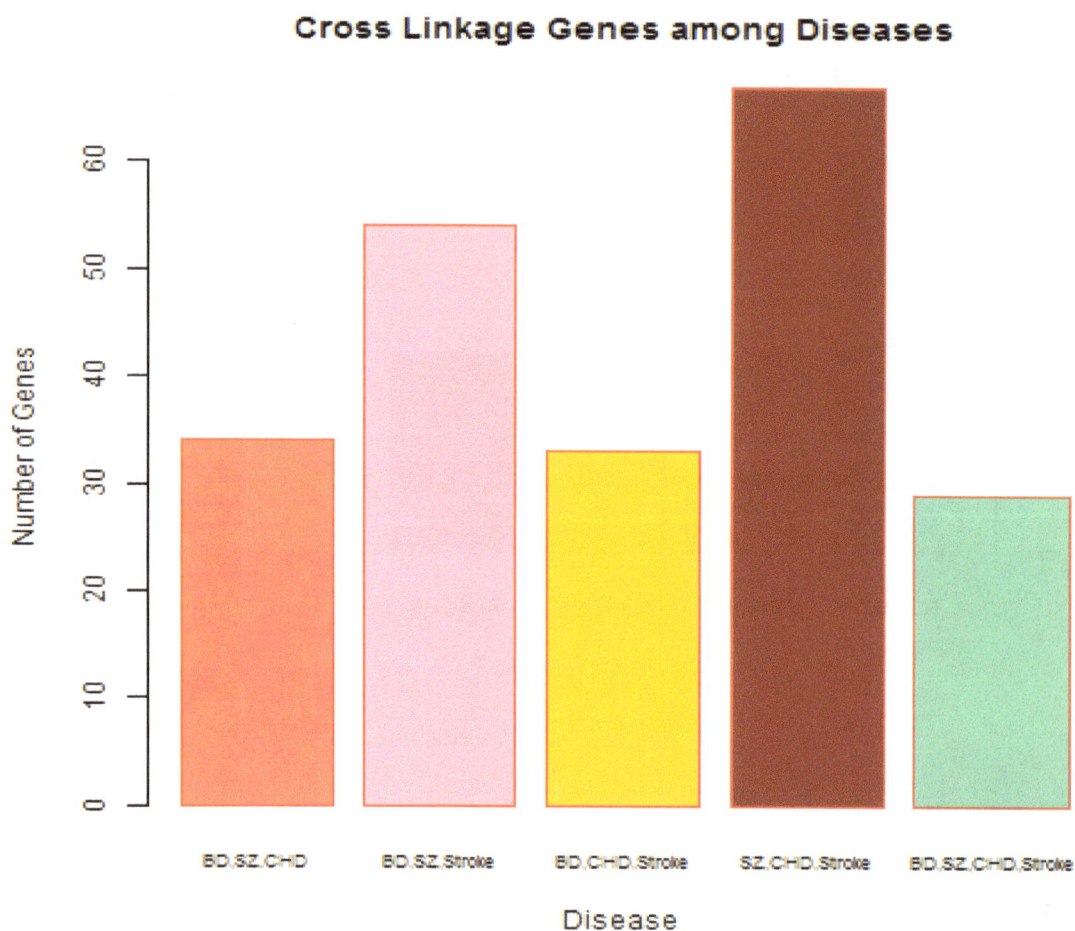

Fig. 3 Cross linkage genes among diseases.

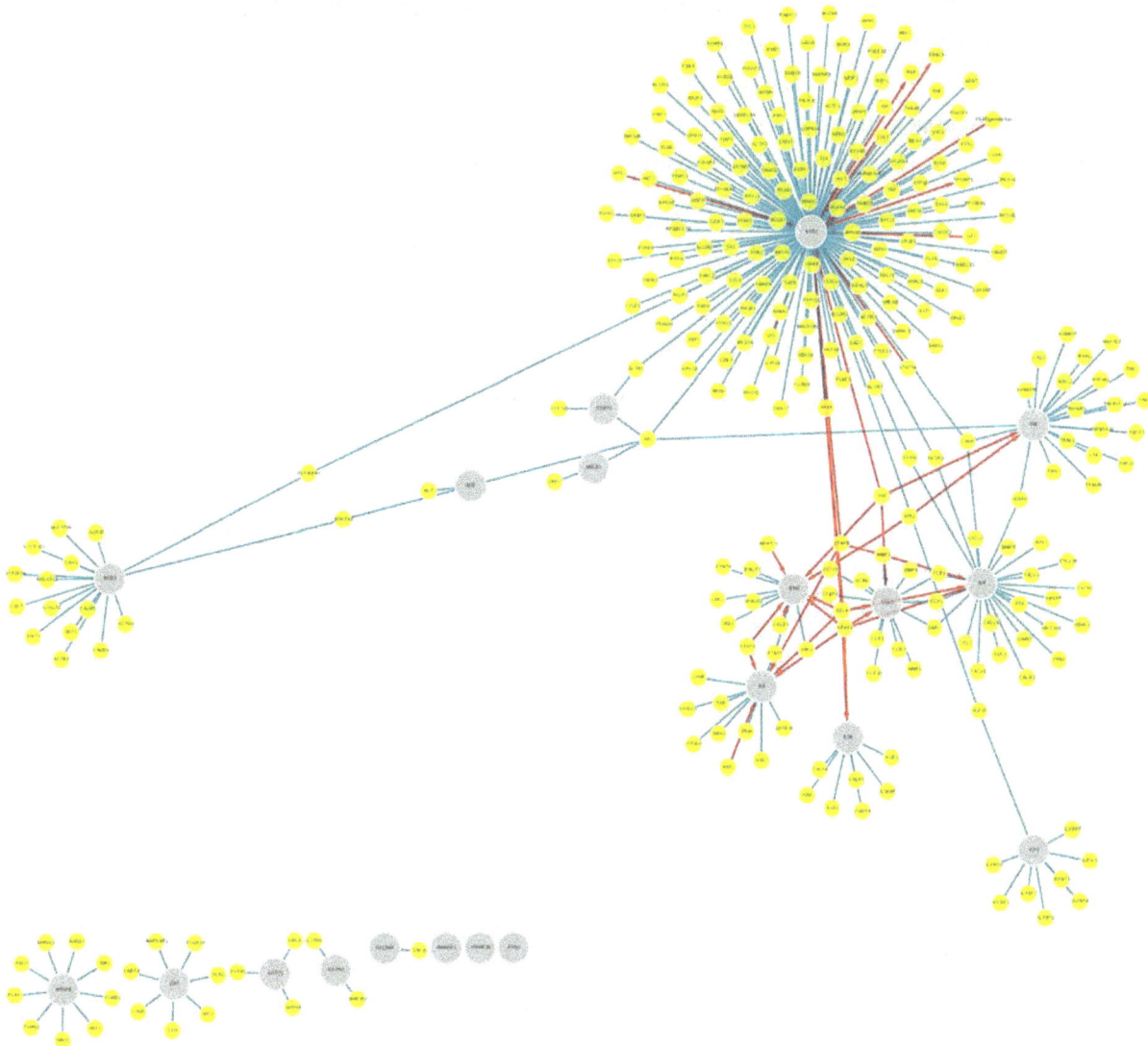

Fig. 4 Protein-protein interaction (PPI) network with 17 common genes.

3.4 Design drug

This is the most vital step of the current research project. A drug is any substance (other than food that provides nutritional support) that, when inhaled, injected, smoked, consumed, absorbed via a patch on the skin, or dissolved under the tongue, causes a physiological change in the body (Wikipedia). In other word, a drug can be defined as a substance used to treat, cure and prevent an illness, relieve from a pain, or modify some specific process in the body for some specific cure. Therapeutic response for a disease is the key root to the invention of a drug. It is the devising process of finding new drugs based on the investigated protein or molecules. The drug should be designed and developed in such a way that it does not disturb the normal chemical process of the body and only affect the target protein. In order to dispose a disease, specific drug needs to be exhibited along with target identification to affect the target genes or proteins. Bioinformatics tools made it easier for researcher to research on specific diseases, leading to the design and development of drug for those diseases. Rapid and revolutionary developments in genome sciences, combinatorial chemistry, informatics and robotics are having major impacts on drug discovery (Blundell, 2002). Protein structure can influence drug discovery at every stage in the design process and can also be used in target identification and selection (Blundell et al., 2006). So, before designing a drug, analyzing or creating PPI Network is important. From the PPI network and common pathway a drug target can be designed using UniHi tool.

4 Results and Discussion

A disordered or any abnormal condition caused by some interrelated genes is called a disease. The interrelated genes are downloaded from trustable gene databases using bioinformatics tools whose are then used to drug design.

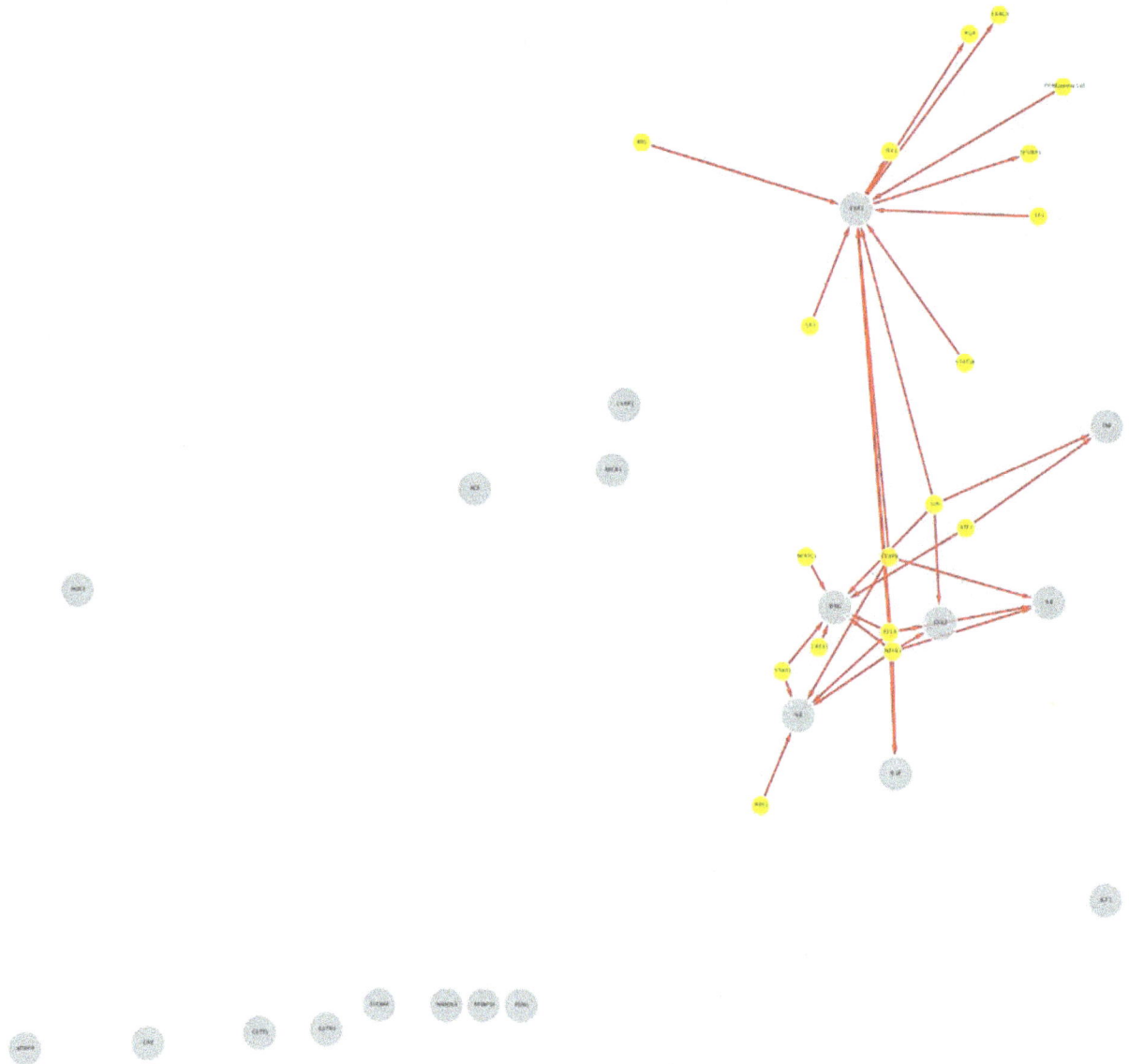

Fig. 5 Regulatory interaction network with 17 common genes.

4.1 Gene filtering

The collected responsible genes without preprocessing & filtering are estimated as 640 for BD, 2555 for schizophrenia, 294 for CHD and 1028 for Stroke and after preprocessing as well as filtering the corresponding genes for *Homo sapiens* are 619 for BD, 1612 for Schizophrenia, 288 for CHD and 638 for stroke (Nahida, 2016). Before and after filtering the number of genes for each specific disease is displayed using a bar plot as shown in Fig. 1.

4.2 Gene mining

After performing cross linkage between and among the investigated corresponding genes, the resultant cross linkage genes between BD & SZ, BD & CHD, BD & stroke, SZ & CHD, CHD & stroke, SZ & stroke are 363, 39, 66, 96, 153, 171 respectively and among BD & SZ & CHD, BD & SZ & stroke, BD & CHD & stroke, SZ & CHD & stroke are 34, 54, 33, 67 respectively (Nahida, 2016). Finally, cross linkage genes among the investigated four diseases (BD, SZ, CHD, stroke) are 29. After mining these 29 common genes and taking the common genes from top 50 weighted genes 17 common genes results. They are TNF, MTHFR, IL6, ACE, ESR1, SLC6A4, ABCB1, CRP, NOS3, IL1B, CXCL8, GSTM1, IGF1, IFNG, GSTT1, PON1 and CCL2. The bar plot in Fig. 2 and Fig. 3 shows the cross linkage genes between and among associated diseases.

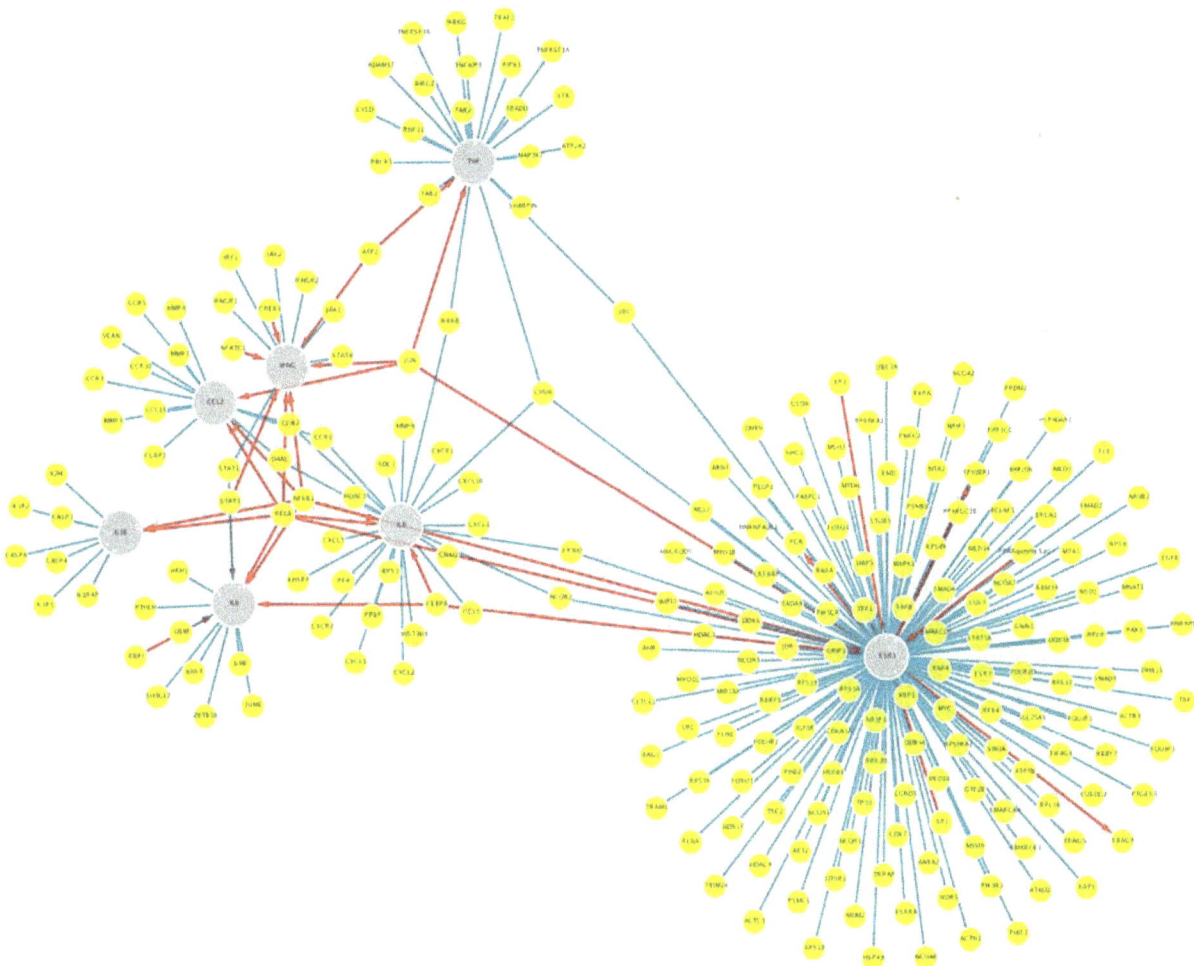

Fig. 6 Protein-protein interaction (PPI) network with 7 common genes.

4.3 PPI network and regulatory interaction

Using UniHI tool, a PPI network and a regulatory interaction network is created. These networks are used to represent the directly or indirectly connected gene and protein interaction. Fig. 4 and Fig. 5 shows the PPI and Regulatory Interaction Network respectively. From Fig. 4 and Fig. 5, it is clear that only 7 genes are responsible for direct interconnection with each other in the networks. They are TNF, IL6, ESR1, IL1B, CXCL8, IFNG, CCL2. So, these common 7 genes are now used to create a PPI Network as displayed in Fig. 6. Regulatory interaction network displays the directly interacted proteins with the target diseases. The regulatory interaction network of Fig. 7 represents the only interacted proteins of associated genes of target diseases.

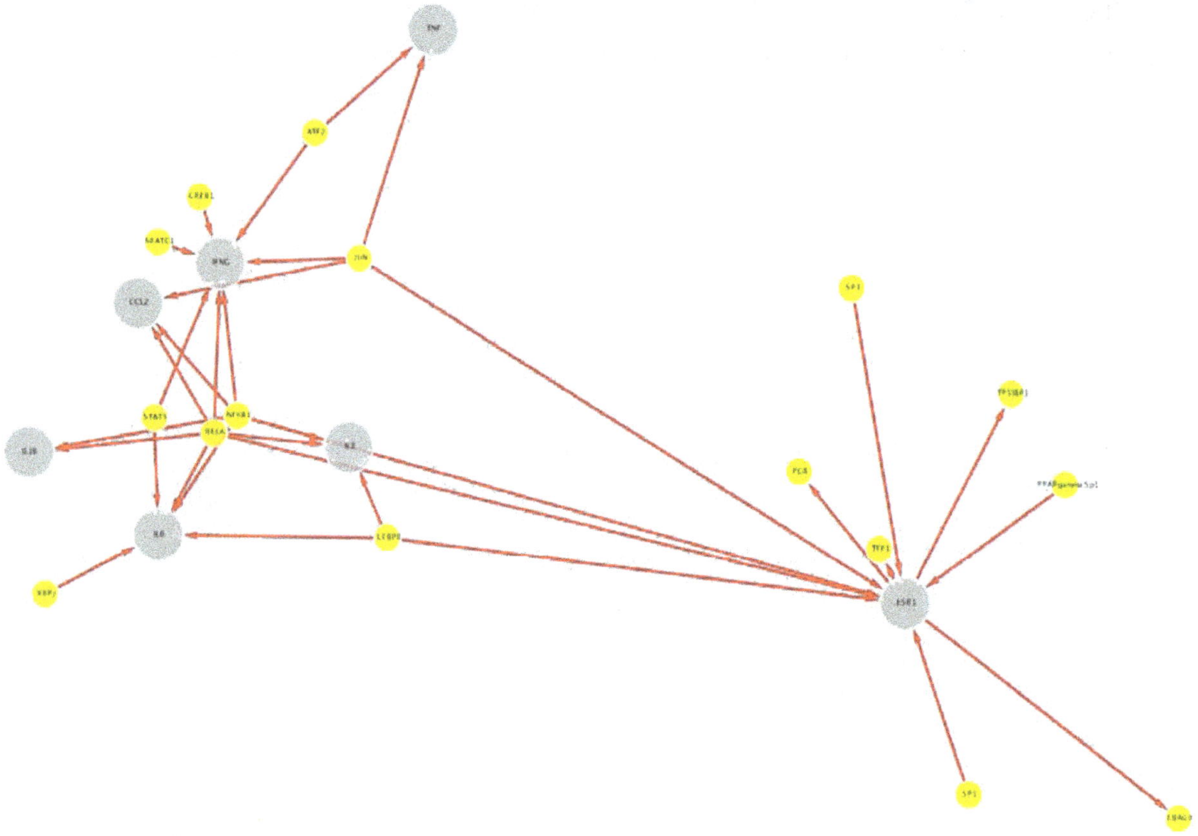

Fig. 7 Regulatory interaction network with 7 common genes.

Cytoscape is an open source software project for integrating biomolecular interaction networks with high-throughput expression data and other molecular states into a unified conceptual framework (Shannon et al., 2003). The network in Fig. 8 is generated in Cytoscape with 7 common genes. STRING (Search Tool for the Retrieval of Interacting Genes/Proteins) is a biological database that contains various information and shows predicted protein–protein interactions. In Fig. 9, the network with 7 common genes is generated using STRING.

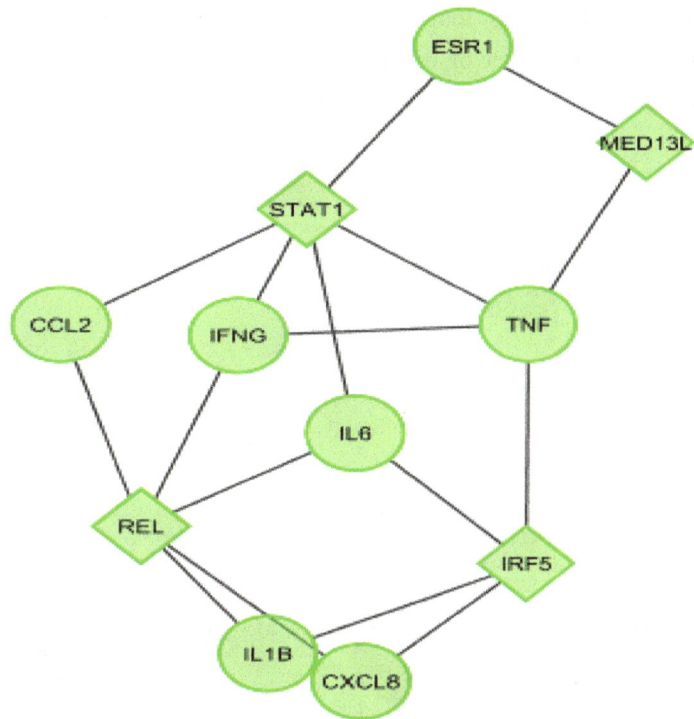

Fig. 8 Network with 7 common genes using Cytoscape.

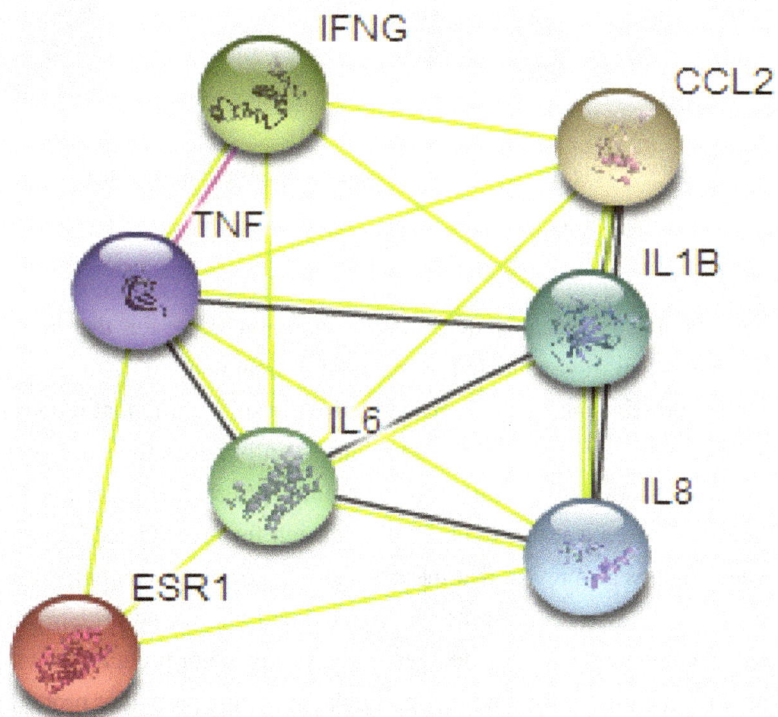

Fig. 9 Network with 7 common genes using String.

GeneMANIA is a bioinformatics tool that shows functional association data, genetic interactions, pathways and co-expression for a set of input data. From the common 7 genes GeneMANIA creates a network as shown in Fig. 10.

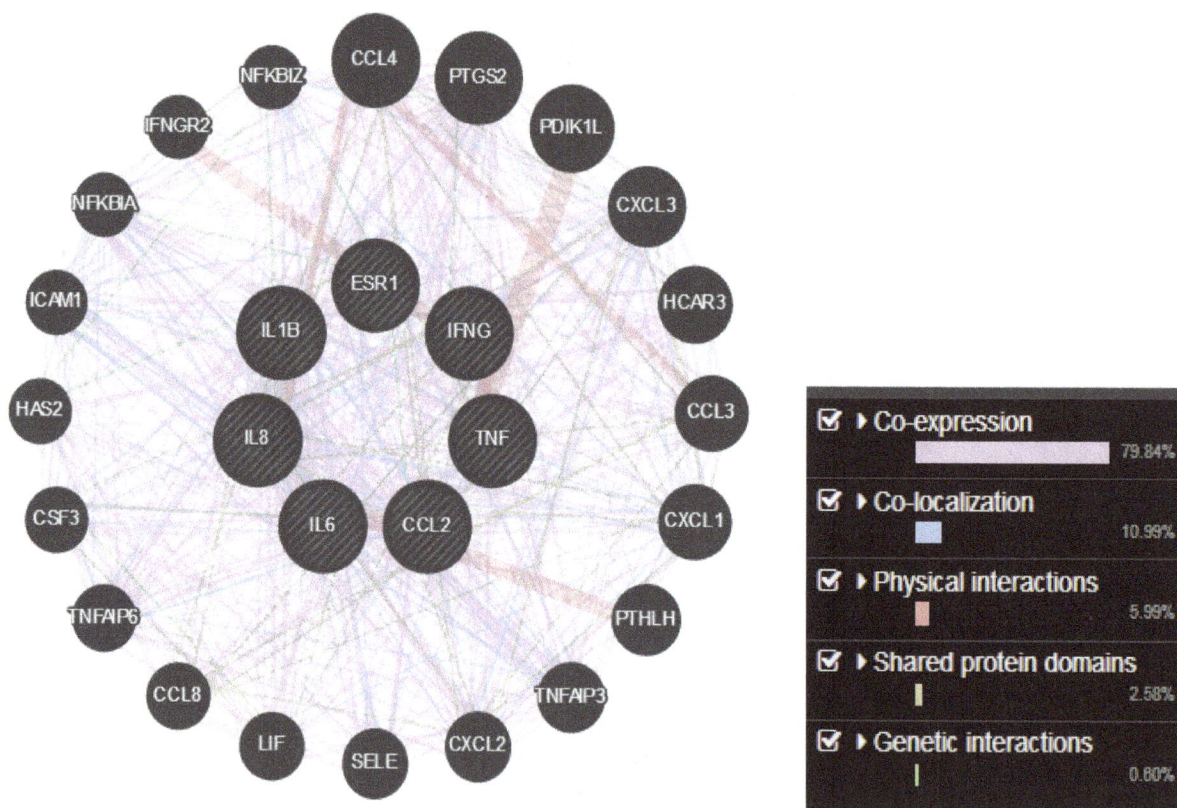

Fig. 10 Network with 7 common genes using GeneMania.

4.4 Design drug

The ultimate goal of this research is designing a common drug for the investigated diseases. Target identification is insufficient for achieving a successful treatment of a disease. Real drugs need to be developed. The target proteins must be influenced by drug in such a way that it does not interfere with normal metabolism. To achieve this activity of protein, various bioinformatics tools are developed. UniHi tool is one of popular bioinformatics tool. Using UniHi tool, a common drug target is designed for selected diseases. Fig. 11 and Fig. 12 are exhibited the structure based drug design for investigated 7 common genes from different side of views. These figures also reveal that there is a strong correlation among 7 investigated genes via some proteins. Figure 11 and 12 also demonstrate the affected and unaffected proteins in the structure. Red color proteins has directed compound with targeted genes and yellow color proteins are vice versa.

It is well-known that a drug must bind to a particular spot on a particular protein or nucleotide. We need to identify and study the lead compounds that have direct activity against a disease. To find out the proteins whose are directly interacted with target diseases, filtering technique is used here. Fig. 13 and 14 illustrate the only interacted proteins of associated genes of target diseases from different point of view. We often apply in several techniques and test a large numbers of compounds from a database that have available structures.

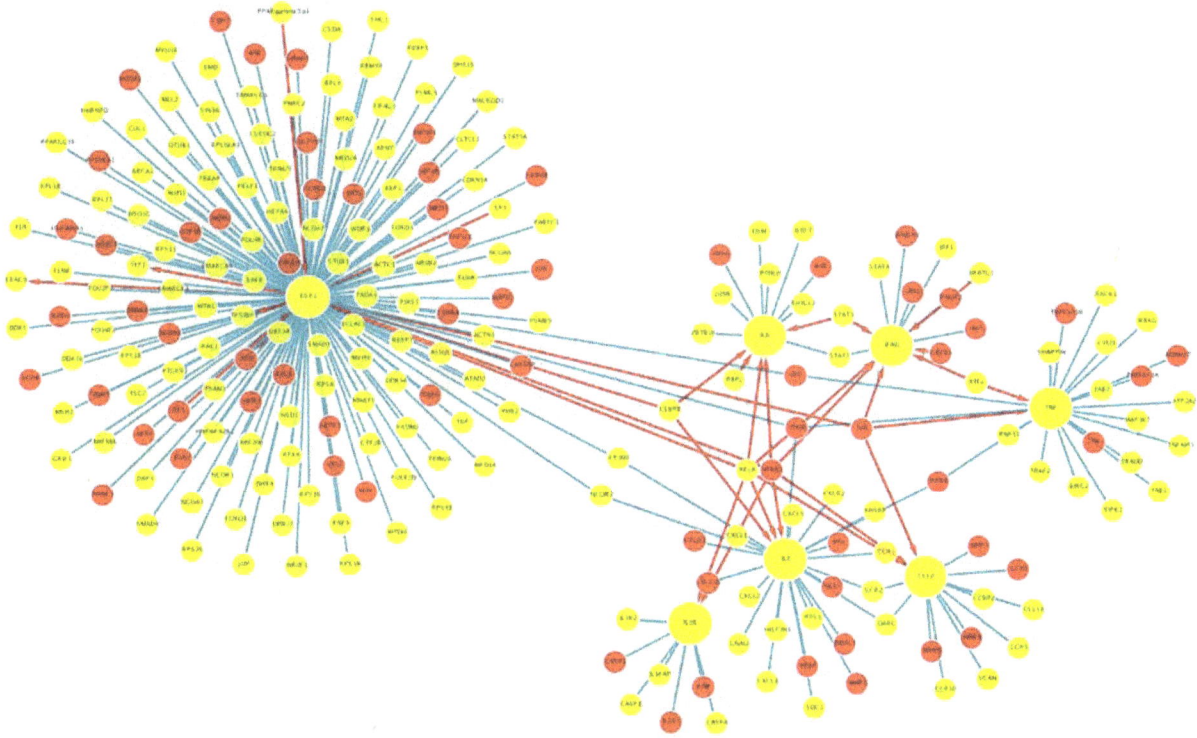

Fig. 11 Drug target network with 7 common genes from side view-1.

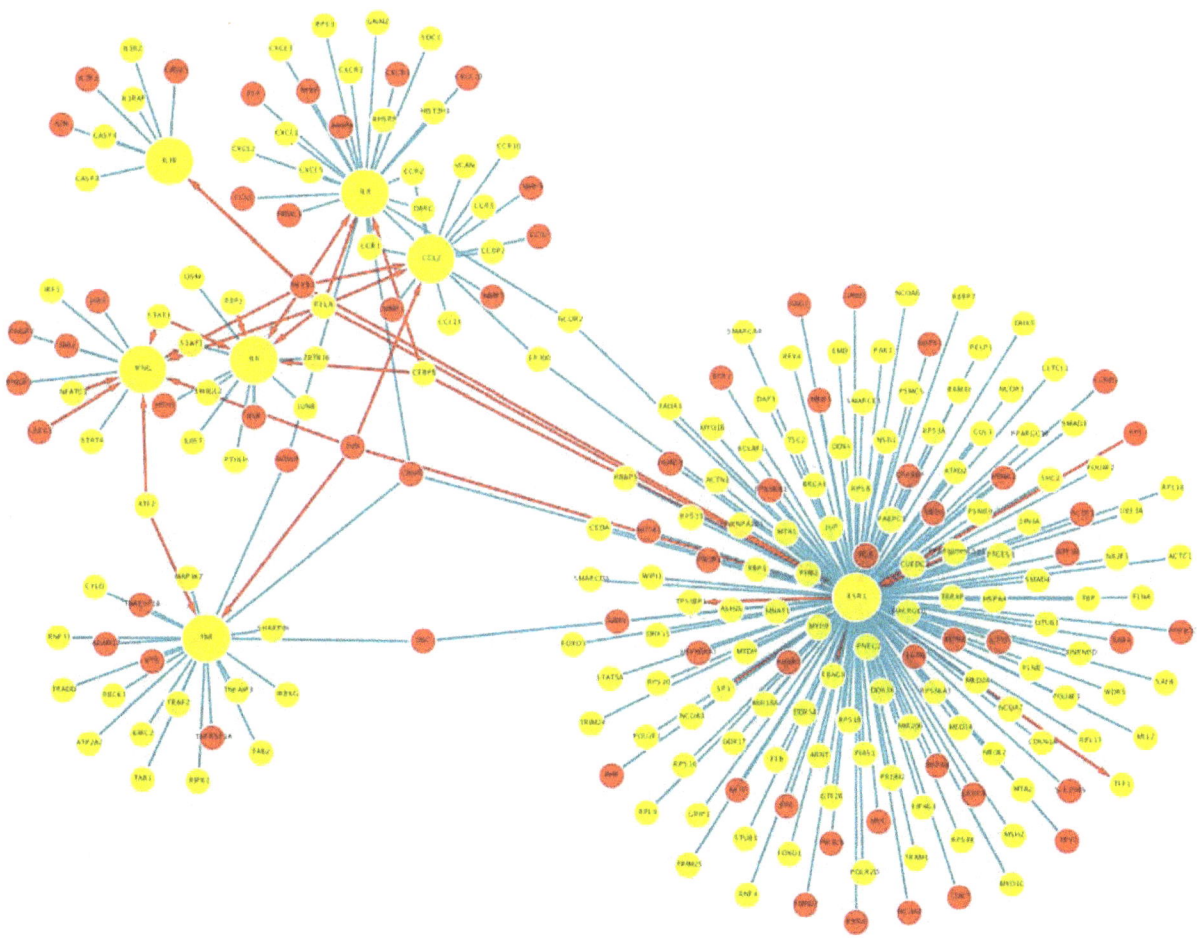

Fig. 12: Drug target with 7 common genes from side view-2.

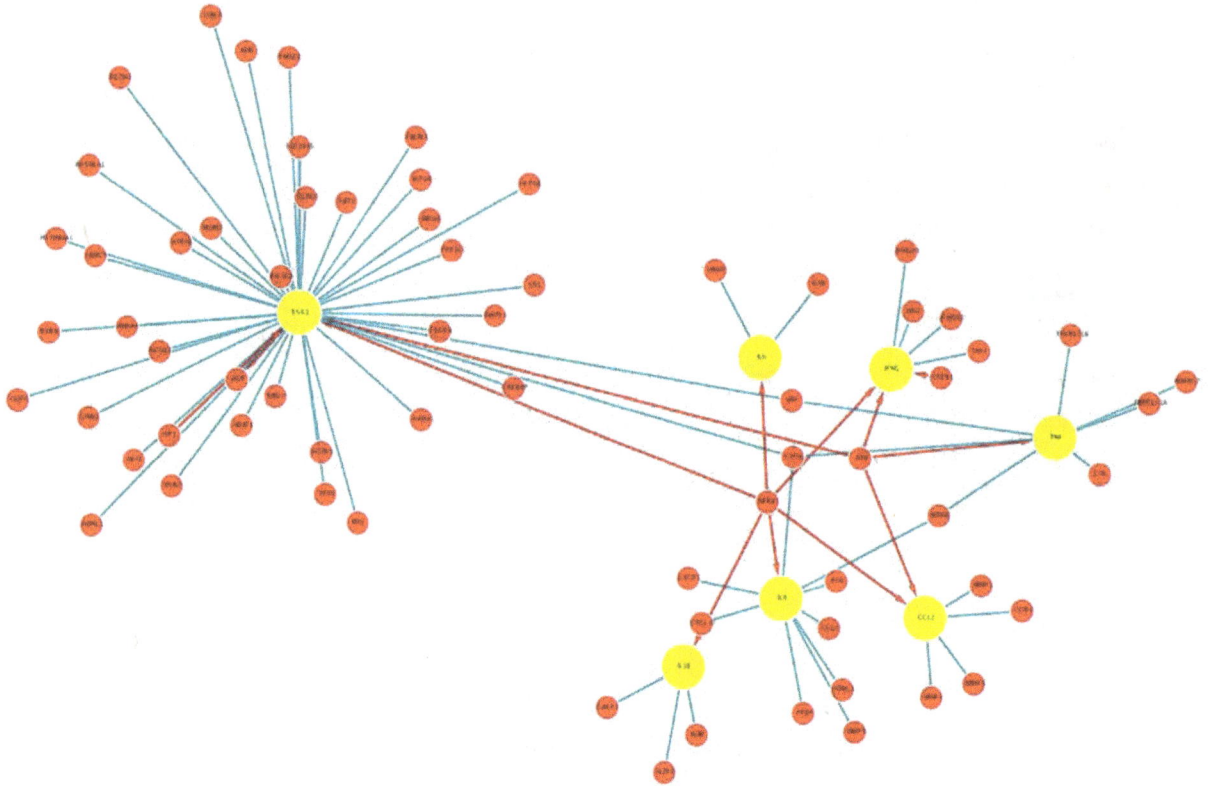

Fig. 13 Drug target network with 7 common genes after filter from side view-1.

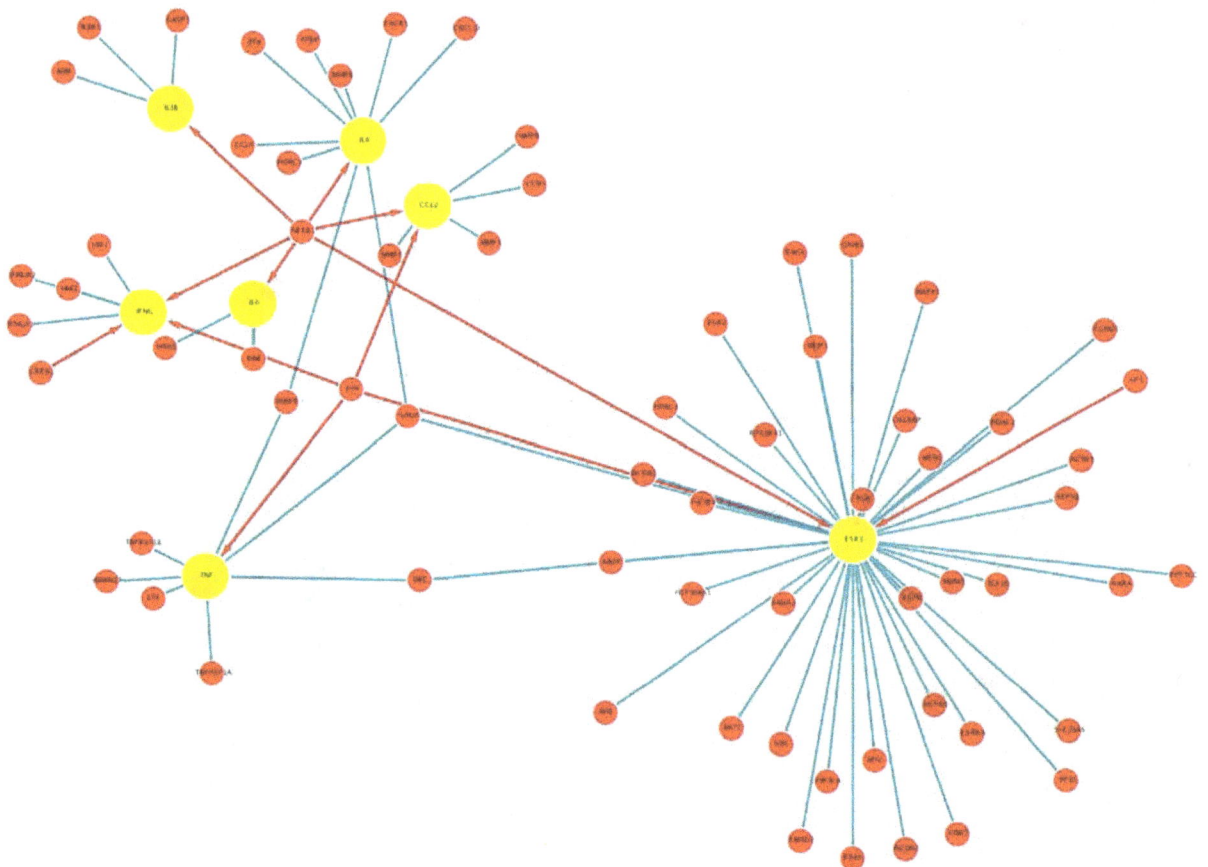

Fig. 14 Drug target network with 7 common genes after filter from side view-2.

5 Conclusions

In the genomic revolution, the contribution of Bioinformatics is incredible. The developments of Bioinformatics tools have disclosed new research area and made uncompromising task easier. Researcher can now analyze diseases, identify and process disease genes to cure the diseases. Computer aided drug design aimed to design a drug to redress certain diseases. Target identification and drug design leads to the development of a drug. According to Anderson (2003), many years of research may be necessary to convert a drug lead into a drug that will be both structure-based drug design and includes, primarily, effective and tolerated by the human body.

Major depression is the key causes of BD, schizophrenia, CHD and stroke diseases. To create a common drug for these diseases the corresponding genes of each disease are studied and particular operations are performed on them. Finally, in this research project a common drug is designed that will remedy these diseases. This research is mainly helpful to understand the PPI, Regulatory Interaction Network and to target drug design. The future work of the research is to work on several other interrelated diseases to design a common drug for those diseases.

Abbreviations

BD= Bipolar Disorder; SZ=Schizophrenia; CHD= Coronary Heart Disease; CVD= Cardiovascular disease.

Acknowledgment

The authors are grateful to the participants who contributed to this research.

References

Akiskal HS, Bourgeois Ml, Angst J, et al. 2000. Re-evaluating the prevalence of and diagnostic composition within the broad clinical spectrumof bipolar disorders. Journal of Affective Disorders, 59: 5-30

Aleman A, Kahn RS, Selten JP. 2003. Sex differences in the risk of schizophrenia: evidence from meta-analysis. Archives of General Psychiatry, 60: 565-571

Anderson AC. 2003. The process of structure-based drug design. Chemistry & Biology, 10: 787-797

Ayalew M, Le-Niculescu H, Levey DF, et al. 2012. Immediate communication: Convergent functional genomics of schizophrenia: from comprehensive understanding to genetic risk prediction. Molecular Psychiatry, 17: 887-905

Baune BT, Adrian I, Arolt V, et al. 2006. Associations between major depression, bipolar disorders, dysthymia and cardiovascular diseases in the general adult population. Psychotherapy and Psychosomatics, 75: 319-326

Belmaker RH. 2004. Bipolar disorder. The New England Journal of Medicine, 351: 476-486

Blundell TL, Jhoti H, Abell C. 2002. High-throughput crystallography for lead discovery in drug design. Nature Reviews Drug Discovery 1: 45-54

Blundell TL, Sibanda BL, Montalva RW, Brewerton S, Chelliah V, Worth CL, Harmer NJ, Davies O, Burke D. 2006. Structural biology and bioinformatics in drug design: opportunities and challenges for target identification and lead discovery. Philosophical Transactions of the Royal Society B, 361: 413-423

Cardno AG, Owen MJ. 2014. Genetic relationships between schizophrenia, bipolar disorder, and schizo affective disorder. Schizophrenia Bulletin, 40(3): 504-515

De Hert et al. 2009. Cardiovascular disease and diabetes in people with severe mental illness position statement from the European Psychiatric Association (EPA), supported by the European Association for the Study of Diabetes (EASD) and the European Society of Cardiology (ESC). European Psychiatry, 24: 412-424

Drug. Wikipedia. 2017. https://en.wikipedia.org/wiki/Drug (last access: 23/04/2017)

Edvardsen J, Torgersen S, Roysamb E, Lygren S, Skre I, Onstad S, et al. 2008. Heritability of bipolar spectrum disorders. Unity or heterogeneity? Journal of Affective Disorders, 106: 229-240

Everson SA., Roberts RE., Goldberg DE, Kaplan GA. 1998. Depressive symptoms and increased risk of stroke mortality over a 29-year period. Archives of Internal Medicine, 158: 1133-1138

Goldstein BI, Carnethon MR, Matthews KA, et al. 2015. Major depressive disorder and bipolar disorder predispose youth to accelerated atherosclerosis and early cardiovascular disease: A scientific statement from the American Heart Association. Circulation, 132: 965-986

Habib N, Ahmed K, Jabin I, Rahman MM. 2016. Application of R to investigate common gene regulatory network pathway among bipolar disorder and associate diseases. Network Biology, 6(4): 86-100

Hoyer EH, Mortensen PB, Olesen AV. 2000. Mortality and causes of death in a total national sample of patients with affective disorders admitted for the first time between 1973 and 1993. British Journal of Psychiatry, 176: 76-82

Jesmin T, Waheed S, Emran AA. 2016. Investigation of common disease regulatory network for metabolic disorders: A bioinformatics approach. Network Biology, 6(1): 28-36

Joukamaa M, Heliovaara M, Knekt P, Aromaa A, Raitasalo R, Lehtinen V. 2001. Mental disorders and cause-specific mortality. British Journal of Psychiatry, 179: 498-502

Kalathur RK, Pinto JP, Hernández-Prieto MA, et al. 2014. UniHI 7: an enhanced database for retrieval and interactive analysis of human molecular interaction networks. Nucleic Acids Research, 42: D408-D414

Keenan NL, Shaw KM. 2011. Coronary heart disease and stroke deaths - United States, 2006. MMWR Surveill Summ, 60(Suppl): 62-66

Klingstrom T, Plwczynski D. 2010. Protein-protein interaction and pathway databases, a graphical review. Briefings in Bioinformatics, 12(6): 702-713

Krishnan KR. 2005. Psychiatric and medical comorbidity of bipolar disorder. Psychosom, 67: 1-8

Li JR, Zhang WJ. 2013. Identification of crucial metabolites/reactions in tumor signaling networks. Network Biology, 3(4): 121-132

Lozano R, Naghavi M, Foreman K, et al. 2012. Global and regional mortality from 235 causes of death for 20 age groups in 1990 and 2010: a systematic analysis for the Global Burden of Disease Study 2010. The Lancet, 380: 2095-2128

McGrath J, Saha S, Welham J, et al. 2004. A systematic review of the incidence of schizophrenia: the distribution of rates and the influence of sex, urbanicity, migrant status and methodology. BMC Medicine, 2: 13

Medaa SA, Ruanob G, Windemuthb A, et al. 2014. Multivariate analysis reveals genetic associations of the resting default mode network in psychotic bipolar disorder and schizophrenia. PNAS, E2066-E2075

Merikangas KR, Jin R, He JP, Kessler RC, Lee S, Sampson NA, et al. 2011. Prevalence and correlates of bipolar spectrum disorder in the world mental health survey initiative. Archives of General Psychiatry, 68: 241-251

National Alliance on Mental Illness. 2017. NAMI.http://www.nami.org/Learn-More/ Mental-Health-Conditions.(Last Access, 4-4-2017)

NIH. 2017. https://medlineplus.gov/mentaldisorders.html.(Last Access 4-4-2017)

Osby U, Brandt L. Correia N, Ekbom A, Sparen P. 2001. Excess mortality in bipolar and unipolar disorder in

Sweden. Arch Gen Psychiatry, 58: 844–850

Rasic D, Hajek T, Alda M, Uher R. 2013. Risk of mental illness in offspring of parents with schizophrenia, bipolar disorder, and major depressive disorder: A meta-analysis of family high-risk studies. Schizophrenia Bulletin, 40(1): 28-38

Roger VL, Go AS, Lloyd-Jones DM, et al. 2012. Heart disease and stroke statistics--2012 update: a report from the American Heart Association. Circulation, 125: e2–e220

Rosamond W, Flegal K, Furie K, et al. 2008. Heart disease and stroke statistics – 2008 update: a report from the American Heart Association Statistics Committee and Stroke Statistics Subcommittee. Circulation, 117(4): E25–E146

Shannon P, Markiel A, Ozier O, Baliga NS, Wang JT, Ramage D, Amin N, Schwikowski B, Ideker T. 2003. Cytoscape: A software environment for integrated models of biomolecular interaction networks. Genome Research, 13: 2498-2504

Sowden GL, Huffman JC. 2009. The impact of mental illness on cardiac outcomes: A review for the cardiologist. International Journal of Cardiology, 132(1): 30-37

TechTarget. 2017. http://searchdatamanagement.techtarget.com/news/2240111413/Bioinformatics-in-Structure Based-Drug-Design. (Last Access 4-4-2017)

Teichmann SA, Babu MM. 2014. Gene regulatory network growth by duplication. Nature Genetics, 36(5): 492-496

Thorup et al. 2015. The Danish high risk and resilience study. BMC Psychiatry, 15: 233

Tsai SY, Lee CH, Kuo CJ, Chen CC. 2005. A retrospective analysis of risk and protective factors for natural death in bipolar disorder. Journal of Clinical Psychiatry, 66: 1586–1591

Weeke A, Juel K, Vaeth M. 1987. Cardiovascular death and manic-depressive psychosis. Journal of Affective Disorders, 13: 287-292

Zhang WJ. 2016a. A mathematical model for dynamics of occurrence probability of missing links in predicted missing link list. Network Pharmacology, 1(4): 86-94

Zhang WJ. 2016b. Network pharmacology: A further description. Network Pharmacology, 1(1): 1-14

Zhang WJ, Feng YT. 2017. Metabolic pathway of non-alcoholic fatty liver disease: Network properties and robustness. Network Pharmacology, 2(1): 1-12

5

A deeper insight into the equilibrium of biological and ecological networks

Alessandro Ferrarini

Department of Evolutionary and Functional Biology, University of Parma, Via G. Saragat 4, I-43100 Parma, Italy

E-mail: sgtpm@libero.it, alessandro.ferrarini@unipr.it, a.ferrarini1972@libero.it

Abstract

The equilibrium of biological and ecological networks is often studied using eigenvector-eigenvalue analyses in order to reckon steady/unsteady properties and trajectories. Although at equilibrium inputs equal outputs for all the system variables, network flows continue to happen. Therefore, in this study I face three underestimated topics of network equilibrium: equilibrium flows, equilibrium sensitivity and equilibrium what-if properties. Using an applicative example, I show here that these three topics add important details to the knowledge of network behaviour at equilibrium.

Keywords equilibrium flows; equilibrium sensitivity; equilibrium what-if; network equilibrium.

1 Introduction

The stability analysis of dynamical networks is a well-established topic, both in ecology and in biology. The equilibrium of biological and ecological networks is often studied using eigenvector-eigenvalue analyses in order to compute their steady/unsteady properties and trajectories. Using a different perspective, Ferrarini (2015) demonstrated that early, or late, stability can be imposed to any kind of ecological and biological network. Ferrarini (2016a) showed that network stability can be imposed also locally, not only globally.

In this study, I face three underestimated topics of network equilibrium: equilibrium flows, equilibrium sensitivity and equilibrium what-if. Using an applicative example, I show that these three topics can add important details to the knowledge of network behaviour at equilibrium.

2 Mathematical Formulation

A generic ecological (or biological) system composed by a set **S** of n interacting taxonomic resolutions (species, genera, family, etc.) or aggregated assemblages of taxa (e.g., phytoplankton) is

$$\frac{d\mathbf{S}}{dt} = \gamma(\mathbf{S}(t))$$

(1)

where S_i is the number of individuals (or the total biomass) of the generic *i-th* taxonomic resolution or assemblage of taxa.

If we also consider time-dependent inputs (e.g. species reintroductions) and outputs (e.g. hunting) from outside, we must write:

$$\frac{d\mathbf{S}}{dt} = \gamma(\mathbf{S}(t)) + \mathbf{I}(t) + \mathbf{O}(t)$$

(2)

with $\mathbf{I}(t) \geq 0$ and $\mathbf{O}(t) \leq 0$, and with initial values

$$\mathbf{S}_0 = <S_1(0), S_2(0)...S_n(0)>$$

(3)

and co-domain limits

$$\begin{cases} S_{1min} \leq S_1(t) \leq S_{1max} \\ ... \qquad\qquad\qquad \forall t \\ S_{nmin} \leq S_n(t) \leq S_{nmax} \end{cases}$$

(4)

Stability happens at time $t = T^{Eq}$ when

$$\begin{cases} \dfrac{dS_1}{dt} = 0 \\ ... \\ \dfrac{dS_n}{dt} = 0 \end{cases}$$

(5)

or, by relaxing Eq. 5, when

$$\begin{cases} \dfrac{dS_1}{dt} \leq \varphi \\ ... \qquad\qquad \text{with } \varphi \to 0 \\ \dfrac{dS_n}{dt} \leq \varphi \end{cases}$$

(6)

2.1 Reckoning network flows at equilibrium

Although at equilibrium inputs equal outputs for all of the variables, network flows continue to happen in the form

$$\begin{cases} \dfrac{dS_1}{dt}^{EQ} = inputs_{S1}{}^{Eq} - outputs_{S1}{}^{Eq} = 0 \\ ... \\ \dfrac{dS_n}{dt}^{EQ} = inputs_{Sn}{}^{Eq} - outputs_{Sn}{}^{Eq} = 0 \end{cases}$$

(7)

thus one main question is the amount of inputs and outputs at equilibrium for each variable, and also their ratios if compared to the equilibrium values. In fact, if flows at equilibrium are high compared to equilibrium values, a small perturbation of the network can lead to great changes, and vice versa.

The flow amount at equilibrium for each generic variable S_i is

$$Inputs_{Si}^{Eq} = \gamma(\mathbf{S}_+^{Eq}(t)) + \mathbf{I}^{Eq}(t) = outputs_{Si}^{Eq} = \left| \gamma(\mathbf{S}_-^{Eq}(t)) + \mathbf{O}^{Eq}(t) \right| \tag{8}$$

where S_+^{Eq} is the set of variables with positive input to S_i, and vice versa for S_-^{Eq}.

For instance, in a simple linear dynamical system composed by species <A, B, C, D> with equilibrium values <a, b, c, d> and interactions upon A equal to <0.3, -0.4, -0.9, 0.6>, then

$$Inputs_A^{Eq} = 0.3 * a + 0.6 * d = outputs_A^{Eq} = \left| -0.4 * b - 0.9 * c \right| \tag{9}$$

Thus, for the generic variable S_i the ratio between equilibrium flow and equilibrium value is

$$\frac{Inputs_{Si}^{Eq}}{Si^{Eq}} = \frac{\gamma(\mathbf{S}_+^{Eq}(t)) + \mathbf{I}^{Eq}(t)}{Si^{Eq}} = \frac{Outputs_{Si}^{Eq}}{Si^{Eq}} = \frac{\left| \gamma(\mathbf{S}_-^{Eq}(t)) + \mathbf{O}^{Eq}(t) \right|}{Si^{Eq}} \tag{10}$$

2.2 Reckoning network what-if at equilibrium

Another important issue is how and how much the network equilibrium values change if one network parameter (let's say p) changes. Network what-if analysis can disentangle the directionality and the degree of network change at equilibrium for a given change of each network parameter. It can be computed as follows

$$\frac{\Delta_{\%} \mathbf{S}^{Eq}}{\Delta_{\%} p} = \frac{\dfrac{\partial \mathbf{S}}{\mathbf{S}^{Eq}}}{\dfrac{dp}{p}} = \frac{\partial \mathbf{S}}{dp} * \frac{p}{\mathbf{S}^{Eq}} = \frac{\partial \mathbf{S}}{dp} * \frac{p}{\mathbf{S}^{Eq}} \tag{11}$$

In case we vary simultaneously two parameters p and q, we must calculate

$$\frac{\Delta_{\%} \mathbf{S}^{Eq}}{\Delta_{\%} (p,q)} = \frac{\dfrac{\partial^2 \mathbf{S}}{\mathbf{S}^{Eq}}}{\dfrac{dp\ dq}{p\ q}} = \frac{\partial^2 \mathbf{S}}{\mathbf{S}^{Eq}} * \frac{p}{dp} * \frac{q}{dq} = \frac{\partial^2 \mathbf{S}}{dp\ dq} \frac{p*q}{\mathbf{S}^{Eq}} \tag{12}$$

and so on for m parameters.

2.3 Reckoning network sensitivity at equilibrium

If we densify what-if analysis at equilibrium by multiple simulated parameter changes, we have a sensitivity analysis. For example, we can vary 1000 times the parameter p by a ±5%. In pseudo code:

$$
\begin{cases}
m = 0 \\
\text{Do} \\
\dfrac{\Delta_\% \mathbf{S}^{Eq}}{\Delta_\% \, p} \\
\text{with } dp \in [0.95*p,\ 1.05*p] \\
m = m + 1 \\
\text{Until } m = 1000
\end{cases}
\tag{13}
$$

3 An applicative Example

Let's consider the classic Lotka-Volterra predator-prey model (Lotka, 1925; Volterra, 1926). The Lotka-Volterra equations are a combination of first-order, non-linear, differential equations widely used to describe the dynamics of biological systems with two species interacting (one as a prey and the other as a predator).

The Lotka-Volterra model makes five assumptions about the environment and the dynamics of the two interacting species: 1) the prey population finds food at any times; 2) the food supply for the predator depends completely on the size of the prey population; 3) the rate of change of each population is proportional to its size; 4) during the interaction, the environment remains unvarying; 5) predators have unbounded appetency. Since differential equations are used, the solution is deterministic and continuous; this means that the generations of both the predator and prey continually overlap.

The nonlinear Lotka-Volterra model with logistic grow of the prey S_1 is a particular case of Eq.1, and it reads as follows

$$
\begin{cases}
\dfrac{dS_1}{dt} = \alpha S_1 (1 - \dfrac{S_1}{\kappa}) - \beta S_1 S_2 \\
\dfrac{dS_2}{dt} = \beta \gamma S_1 S_2 - \delta S_2
\end{cases}
\tag{14}
$$

with initial values

$$
\vec{S}_0 = <S_1(0),\ S_2(0)>
\tag{15}
$$

and co-domain limits

$$
\begin{cases}
S_{1min} \leq S_1(t) \leq S_{1max} \\
S_{2min} \leq S_2(t) \leq S_{2max}
\end{cases}
\quad \forall t
\tag{16}
$$

Let's consider the Lotka-Volterra system of Eq. 14 with the following parameters and constants:

$$
\begin{cases}
S_1(0) = 100 \\
S_2(0) = 20 \\
\alpha = 4 \\
\beta = 0.05 \\
\gamma = 1 \\
\delta = 4 \\
\kappa = 500 \\
dt = 0.01 \\
\varphi = 0.0001
\end{cases}
$$

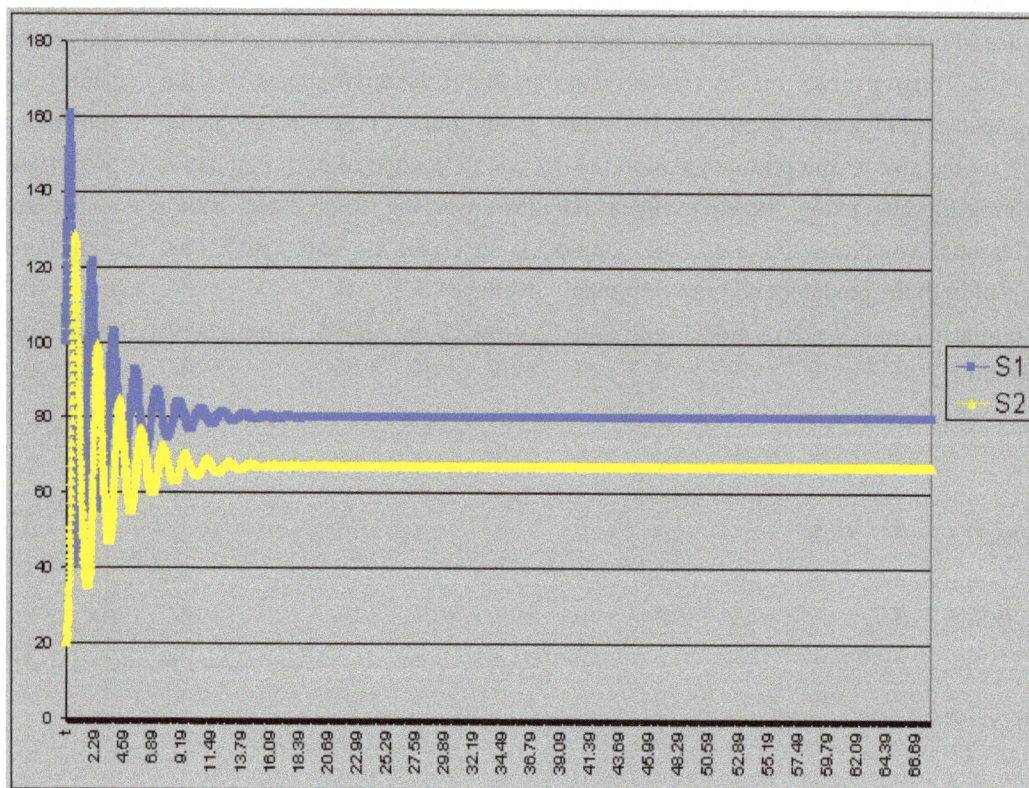

Fig. 1 Time plot of the nonlinear Lotka-Volterra system under study.

This network has eigenvalues $< -0.32 + 3.65i, -0.32 - 3.65i >$ which means that the equilibrium point is stable (both the real parts of the two eigenvalues are negative) with oscillations that reduce the amplitude while nearing the steady point (the sum of imaginary parts is 0). The previous system goes at the steady state with $S_1 = 80.00$ and $S_2 = 67.20$ (Fig. 1).

At equilibrium, the previous biological network has the following flows computed through Eq. 8:

$$Inputs_{S1}{}^{Eq} = 4*80*(1-80/500) = 268.8$$

$$Outputs_{S1}{}^{Eq} = -0.05*80*67.2 = -268.8$$

$$Inputs_{S2}{}^{Eq} = 0.05*80*67.2 = 268.8$$

$$Outputs_{S2}{}^{Eq} = 4*67.2 = -268.8$$

Thus, the ratios between flows and values at equilibrium are

$$\begin{cases} \dfrac{Inputs_{S1}{}^{Eq}}{S1^{Eq}} = \dfrac{268.8}{80} = 3.36 \\ \dfrac{Inputs_{S2}{}^{Eq}}{S2^{Eq}} = \dfrac{268.8}{67.2} = 4 \end{cases}$$

A two-parameter what-if analysis with $\alpha = 4.1$ (instead of 4) and $\delta = 3.8$ (instead of 4), imposed exactly at network equilibrium T^{Eq}, produces the changes in network dynamics illustrated in Fig. 2 and Fig. 3 and computed using Eq. 11. It is very interesting to note the evident shift in the network equilibrium due to the post-equilibrium change in the α and δ parameters. All the analyses have been carried out using the software Control-Lab 8 (Ferrarini, 2016b).

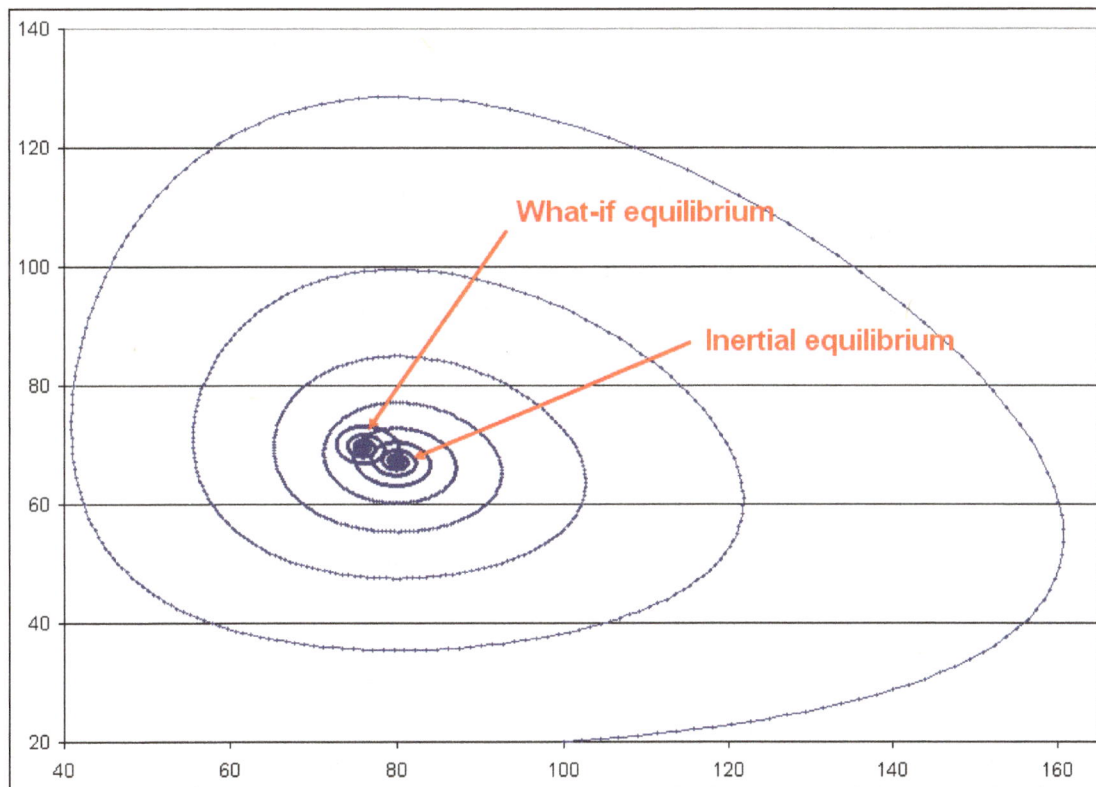

Fig. 2 Two-parameter what-if analysis at equilibrium of the Lotka-Volterra system under study (phase plot).

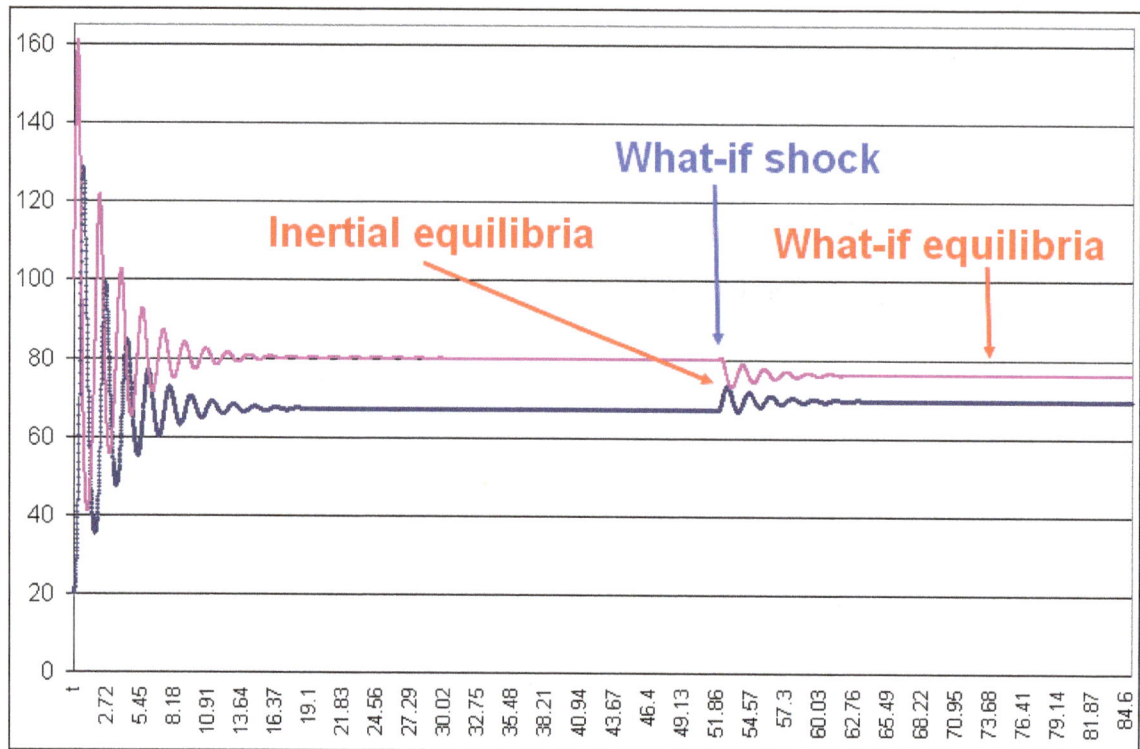

Fig. 3 Two-parameter what-if analysis at equilibrium of the Lotka-Volterra system under study (time plot).

4 Conclusions

The equilibrium of dynamical networks is a well-studied topic both in ecology and in biology. In this paper, I have adopted a different perspective: instead of analyzing the stability of a generic biological or ecological network, I have focused on three underrated topics: equilibrium flows, equilibrium what-if analysis and equilibrium sensitivity analysis.

I have showed that these three topics provide useful details about the way a biological or ecological network behaves, thus providing a deeper perspective into the equilibrium of biological and ecological networks.

References

Ferrarini A. 2015. Imposing early stability to ecological and biological networks through Evolutionary Network Control. Proceedings of the International Academy of Ecology and Environmental Sciences, 5(1): 49-56

Ferrarini A. 2016a. Bit by bit control of nonlinear ecological and biological networks using Evolutionary Network Control. Network Biology, 6(2): 47-54

Ferrarini A. 2016b. Control-Lab 8: a software for the application of Ecological Network Control. Manual, 106 pages

Lotka A.J. 1925. Elements of Physical Biology. Williams & Wilkins Co, Baltimore, USA

Volterra V. 1926. Variazioni e fluttuazioni del numero d'individui in specie animali conviventi. Mem. R. Accad. Naz. dei Lincei. Ser. VI, vol. 2

6

Fibrillar organization in tendons: A pattern revealed by percolation characteristics of the respective geometric network

Daniel Andrés Dos Santos[1], **María Laura Ponssa**[2], **María José Tulli**[2], **Virginia Abdala**[1,2,3]

[1]Instituto de Biodiversidad Neotropical, Facultad de Ciencias Naturales e Instituto Miguel Lillo, Universidad Nacional de Tucumán – CONICET. Horco Molle S/N, Yerba Buena, Tucumán, Argentina

[2]Instituto de Herpetología, Fundación Miguel Lillo-CONICET. Miguel Lillo 251, San Miguel de Tucumán, Tucumán, Argentina

[3]Cátedra de Biología General, Facultad de Ciencias Naturales e IML, Universidad Nacional de Tucumán. Miguel Lillo 251, San Miguel de Tucumán, Tucumán, Argentina

E-mail: dadossantos@csnat.unt.edu.ar

Abstract

Since the tendon is composed by collagen fibrils of various sizes connected between them through molecular cross-links, it sounds logical to model it via a heterogeneous network of fibrils. Using cross sectional images, that network is operatively inferred from the respective Gabriel graph of the fibril mass centers. We focus on network percolation characteristics under an ordered activation of fibrils (progressive recruitment going from the smallest to the largest fibril). Analyses of percolation were carried out on a repository of images of digital flexor tendons obtained from samples of lizards and frogs. Observed percolation thresholds were compared against values derived from hypothetical scenarios of random activation of nodes. Strikingly, we found a significant delay for the occurrence of percolation in actual data. We interpret this finding as the consequence of some non-random packing of fibrillar units into a size-constrained geometric pattern. We erect an ideal geometric model of balanced interspersion of polymorphic units that accounts for the delayed percolating instance. We also address the circumstance of being percolation curves mirrored by the empirical curves of stress-strain obtained from the same studied tendons. By virtue of this isomorphism, we hypothesize that the inflection points of both curves are different quantitative manifestations of a common transitional process during mechanical load transference.

Keywords percolation; collagen; fibril network; interspersion; pattern recognition.

1 Introduction

It has been often stressed that collagen fibrils in a connective tissue exhibit a network organization (Purslow et al., 1998; Berthod et al., 2001; Chandran and Barocas, 2006; Rigozzi et al., 2011; Shirazi et al., 2011). It is basically proposed that the microstructural properties of the collagen network contribute to continuum mechanical tissue properties that are strongly anisotropic with tensile-compressive asymmetry (Shirazi et al., 2011). Besides, extensible connective tissues (e.g. skin, blood vessels, fascia) contain networks of fibrillar collagen embedded into an amorphous matrix. It is the reorientation of the collagen fibres within these networks that allows large extensions of the tissues and is responsible for their non-linear stress–strain curves (Wainwright et al., 1976; Purslow et al., 1998). This characterization of the collagen organization as a network is used to explain mechanical properties of soft biological tissues. Despite distinct mechanical functions, biological soft tissues have a common microstructure in which a ground matrix is reinforced by a collagen fibril network (Purslow et al., 1998). Likewise, the existence of a collagen network in hard biological tissues such as cartilage (Långsjö et al., 2009; Långsjö et al., 2010; Julkunen et al., 2010) is also widely accepted. Collagen fibrils are not isolated functional entities, but they integrate a network system in which the proximate fibrils could exhibit a functional connection. In spite of the currently assumption that the collagen fibril network is responsible of the main mechanical properties of tendon, to our best knowledge the underlying geometrical network has been never formalized in graph terms where the nodes and edges represent the fibrils and their cross-links respectively.

The properties of connective tissues are known to depend on a wide variety of factors such as the type and maturity of the tissue, the chemical nature of the covalent cross-links, the type and quantity of the glycosaminoglycans throughout the extracellular matrix (ECM) and the content of elastic fibres, water and minerals (Parry et al., 1978).Two main classes of extracellular macromolecules make up the tendon matrix: proteoglycans (PGs), which play a complex role in force transmission and maintenance of tendon tissue structure (Reed and Iozzo, 2002; Rigozzi et al., 2010) and collagen fibrils. Studies that have investigated the relationship between structural and mechanical properties have generally focused on one major component, either collagen or proteoglycans (PG), with studies focusing on collagen fibril morphology being more common (Rigozzi et al., 2010).The amount of interactions between the collagen fibrils and the surrounding matrix influents the stiffness of the tissue, and this may prevent changes in shape after the removal of stress (Parry et al., 1978). The degree of interaction between the collagen fibrils and the amorphous matrix is function of the collagen fibril diameter distribution. The relevant role of the ECM is also visible during tendon development, because collagen fibrillogenesis generates a tendon-specific extracellular matrix that determines the functional properties of the tissue (Zhang, 2005; Zhang et al., 2005). Other studies indicate that biochemical deficiencies the amorphous ECM may be a primary causative factor in certain tendon pathologies (Battaglia et al., 2003; Mikic et al., 2001).

Networks are a collection of elements (nodes or vertices) connected by some relationships of interest (links or edges) (Zhang, 2012a, 2012b, 2012c). The internet, airline routes, and electric power grids are all examples of networks whose function relies crucially on the pattern of interconnection between the components of the system. Thinking of systems as networks and studying their patterns of connection can often lead to new and useful insights (Newman, 2010; Ferrarini, 2013, 2014; Zhang, 2012a, 2012b, 2012c, 2013). An important property of such connection patterns is their robustness—or lack thereof— to removal of network nodes, which can be modeled as a percolation process on a graph representing the network (Callaway et al., 2000). Percolation theory is a branch of probability theory dealing with properties of random media (Berkowitz and Ewing, 1998; Zhang, 2012a, 2013). It is one of the simplest models in probability theory which exhibits what is known as critical phenomenon. This usually means that there is a natural parameter in the model at which the behavior of the system drastically changes (Grimmett, 1999). Percolation statement is simple: every site on a specified lattice is independently either occupied (recruited), with probability q, or not with probability $1 - q$. In particular, the system percolates when it exhibits a continuous phase transition at a finite value of q which, on a regular lattice, is characterized by the formation of a cluster large enough to span the entire system across its dimensions in the limit of infinite system size (Newman and Ziff, 2001). From a mathematical perspective, the percolation theory describes the behavior of connected clusters in a random graph (Cuestas et al., 2011). The percolation method allows us to evaluate the network resilience to deletion of network nodes (Callaway et al., 2000).

In this work we delineate a network approach for studying the tendon organization (system composed of interconnected fibrils) that could contribute to a better understanding of its biomechanical responses. Relative neighborhood networks between collagen fibrils are here erected as proxies for the underlying collagen network. This type of objects constitutes an appropriate candidate to make mathematically tractable the collagen network. The links (edges) between the discrete units (nodes) of this network are derived from the relationships of spatial proximity between fibrils. However, the spatial gap between fibrils connected by proximity is occupied by the amorphous matrix and should be also considered a relevant component for the functionality of the entire system. Amorphous matrix is assumed to be the physical substrate over which the information can flow across the nodes of the network. In dealing with the term transference of information, we adopt the meaning implicit to the information theory (Shannon, 1948) that involves the transmission of data or any state change in a system. This notion liberates us from considering the ECM network as a theoretical model uniquely associated to the context of force transmission. The model of ECM network is useful to address the topic of functional integrity. One way to do that is to evaluate if information can propagate throughout the structure of the network, or equivalently to study its characteristics of percolation. Percolation theory may contribute to the understanding of the information flow such as force transmission from the beginning of the tensile activity (tendon activated by a contractile force) until the resulting response (movement of skeletal pieces).

This paper is interested on the morphology and spatial organization of the tendon collagen fibrils in addition to their functional consequences. It is structured around the following assumptions: i) collagen fibrils are units that mediate the transmission of information; ii) the tendon is an assembly of interconnected fibrils that can be modeled with the approach of geometric networks; iii) molecular cross-links provides the material evidence about the connection between fibrils; iv) percolation of information throughout the network is directly associated to the notion of functional integrity; v) phase transition involved by the percolation threshold can be traced to a point on the non-linear stress-strain curve. In tight correspondence with the above premises, and considering a hypothetical scenario of fibril recruitment by increasing size, we ask the following questions: 1) which is the critical threshold that allows the information to traverse the physical dimensions of the network in which fibrils reside? Based on the observed percolation, 2) is it possible to infer some peculiar geometrical feature of the network structure? To answer this, we need to compare the observed percolation pattern against the random expectations under a stochastic shuffling of the fibril sizes but maintaining the topology of the network. Taking into account that percolation of a system implies a phase transition, 3) which is the influence of this putative phenomenon in the mechanical properties exhibited by tendons? Can the stress-strain curves be used to deal with the last issue? We think that a right comprehension of all these inquiries will bring us new insights to grasp the outstanding mechanical properties of the tendon.

2 Material and Methods

Electron microscopy analysis was conducted with samples of the flexor tendon of the Digit IV obtained from adults of anurans (*Scinax nasicus*, *Phyllomedusa sauvagii*, *Rhinella arenarum*, *Leptodactylus latinasus* and *Leptodactylus chaquensis*) and squamatan reptiles (*Liolaemus elongates*, *L. coeruleus*, *L. bibroni* and *Tupinambis rufescens*). Details about examined specimens are provided in the Appendix. We selected the flexor tendon of Digit IV because of the importance of this digit in locomotion as stressed by Teixeira-Filho et al. (2001) and Tulli et al. (2011).

Samples were placed overnight in 0.1-M phosphate buffer with 2.5% glutaraldehyde and 4% paraformaldehyde. The tissue was then fixed in 1% osmium tetroxide, dehydrated in graded acetones, and flat embedded in Epon plastic 812 (Ernest F Fullam, Inc, Latham, NY) in a cross-sectional orientation. Sections (85 nm) were obtained and stained with 0.25% lead citrate and 5% uranyl acetate in 50% acetone and then observed and photographed in a JEOL100CX transmission electron microscope (LAMENOA, Universidad Nacional de Tucumán, Argentina). Collagen fibril diameters of each species were measured on each micrograph using the Image J 1.44p (Wayne Rasband, National Institutes of Health, USA,http://rsbweb.nih.gov/ij/). Diameter of each fibril present in the selected area was measured. Each fibril included in the selected area was identified in a coordinate system using the particle analysis option of the Image J software.

2.1 The collagen fibril network

Each cross section of a tendon can be represented as a set of points distributed in a 2-dimensional space. Since these points are assumed to interchange information they give rise to a spatial network. In fact, glycosaminglycancross-links connect each fibril with the adjacent ones. To address the problem of the geometrical organization of the network, we need to operationalize the concept of the fibrillar network. Because spatial networks subsume into the category of geometric graphs, we choose this mathematical model to work with. A geometric graph $G = (V, E)$ is an embedding of the set V of nodes as points in the plane, and the set E of edges as straight line segments joining pairs of points in V. Locations for the center mass of each fibril are adopted as point occurrences of fibrils throughout the study system (here, the cross section of the tendon). We also say that the graph G is planar if no two of its edges intersect except perhaps at their end points. In the computational geometry literature, there are several classes of planar geometric graphs arising from what are known as proximity graphs (see the survey of Jaromczyk and Toussaint 1992 for more details).

In this paper, we will model the topology of the collagen fibril network via a Gabriel graph. The rationale for this choice relies on the capacity of the Gabriel graph to capture a fair representation of the proximity structure portrayed by the data. A geometric graph $G= (V, E)$ is called a Gabriel graph if the following condition holds: for any $u, v \in V$, an edge $(u, v) \in E$ if and only if the circle with uv as diameter does not contain any other point of V (Gabriel and Sokal, 1969) (Fig. 1). The ultimate meaning of observing two adjacent fibrils A and B in the underlying Gabriel graph (connected by an edge) is that their area of reciprocal influence is unique and not interfered by nearby fibrils. If the stress force does flow throughout the interfibrillar matrix, it would freely flow across the space bridging fibrils directly linked in the respective Gabriel graph.

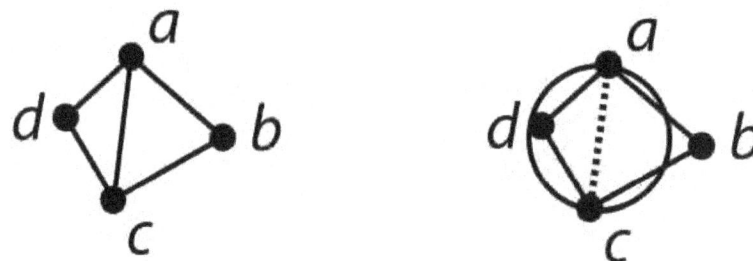

Fig. 1 Example of Gabriel graph (right) obtained from the Delaunay triangulation (left) applied on a set of 4 points, i.e. {a, b, c and d}. The edge joining a to c is discarded because it corresponds to the diameter of a circle that includes another point of the set (i.e. point d).

2.2 Percolation

The network of collagen fibrils is used to represent the propagation of information across the tendon. Here, information can be understood as the propagation of the mechanical stress across the tendon, or the flow of the cross-linking molecules through the amorphous matrix, or any other factor that can change the integral response of the tendon. Vertices on the graph are considered occupied or not, depending on whether the network nodes they represent (fibrils) are activated or recruited. We have examined the site percolation of the network in which the occupation is a function of the fibril size. So, we have considered percolation under a targeted activation of nodes. In this way, we have simulated a sequential recruitment of fibrils increasingly ordered by their sizes and checked then the spanning area of the putative percolating cluster.

In site percolation of spatial networks one views the clusters formed by the occupied vertices. A cluster is defined as set of neighboring or adjacent vertices that are occupied. Of particular interest is to establish if there exists a cluster that connects the borders of the area spanned by the network (here, the cross section of the tendon). Such a cluster is called a percolating cluster and its presence represents a qualitative change in the structure of the system from a disconnected state to a connected one (functionality transition). We estimate the percolation threshold (q) from the inflection point of the S-shaped curved relating the relative size of the spanning cluster and the fraction of activated sites. The following fitting model was used (Pawłowska and Sikorski, 2013):

$$P(\varphi) = 1 - \left(1 + \exp\left(\frac{\varphi - q}{a}\right)\right)^{-1}$$

where a is a parameter that dictates the slope of the curve. Here, percolation probability P is function of the fraction φ of activated sites. In operative terms, nonlinear least square regression was applied on the values of relative size of the percolating cluster against the respective fraction of activated sites. The relative size of the percolating cluster was taken as the ratio between the area of the bounding box enclosing the points of the largest component and the overall rectangular area of the system under study. Percolation threshold was calculated for the case of targeted sequential activation of nodes ordered by increasing size. In a next instance of analysis, this observed score for q was compared against values obtained from random scenarios of node activation. Certainly, if the observed value would deviates from random expectations then we could suspect some pattern of spatial organization where the size of fibrils has a prominent role.

We conducted separate recruitment scenarios for each empirical case of study. Each scenario consisted of 21 instances in a gradual sequence of node activation, using for that purpose a vector of cutoff values that dictate the activation (or not) of the nodes. Whenever a node has a size (= fibril diameter) lower or equal than a specified cutoff value it is activated. The 21 elements of the referred vector are increasingly ordered and correspond to the percentiles by steps of five (0, 5, 10, ..., 100) found on the statistical distribution of fibril sizes. Thus, the first element of this vector corresponds to the minimum size of fibril, its second element corresponds to the fifth percentile, its third element is the tenth percentile, and so on. The last element concerns to the largest fibril. Although exploring the entire set of unique values of size would have yielded more accurate results, we have binned the data by percentile intervals of five because of computational facilities. The full activation of nodes translates into the original Gabriel graph that accounts for the collagen fibril network. As the activation of nodes proceeds from the minimum cutoff value to the maximum one, nodes are connected by an edge if they are actually neighbors in the underlying Gabriel graph. The critical instance, or scene, of the recruitment scenario is that where the percolating cluster firstly appears. The inflection point on the percolation curve, obtained by adjustment of the above equation to the 21 instances of node activation, was used to estimate the percolation threshold q. The q quantile for the set of fibril diameters was subsequently employed for inferring the critical size for fibril activation around which percolation occurs.

To test the significance of the observed values of percolation, we compared them with scores obtained via random simulations. We conserved the network topology and changed the original assignment of size values among the nodes. We set the level of significance at 5% over a total of 100 random simulations. Rejection of null hypothesis would suggest a geometrical pattern shaped by the size of the collagen fibrils.

2.3 Balanced interspersion of heteromorphic fibrils

We will consider a model of non-random interspersion to explain the particular arrangement of polymorphic fibrils, i.e. a pattern where a given fibril is surrounded by others of dissimilar size. We will study an algorithmic arrangement of fibrils on a square lattice able to mimic the percolating behavior observed across the empirical cases of study.

2.4 Stress-strain relationship and inflection point

Tendon samples of *Leptodactylus chaquensis*, *Phyllomedusa sauvagii*, *Rhinella arenarum* and *Tupinambis rufescens* were carefully aligned and gripped using Instron® at room temperature located at Facultad de Odontología, Universidad Nacional de Tucumán. Tendons were tested in tension up to break at the same elongation rate of 0.033 mm/s. Initial gauge length and width of specimens were measured before performing the tensile test. Once the ultimate tensile strength was achieved, we decided to report any additional reading of stress-strain above the 90% of this maximum caused by the gradual breakage of tendons by defibrilation.

Curves were generated through piecewise-cubic splines on the raw data. The inflection point of a curve is the geometrical place where a change of concavity is produced, and they are linked with phase transitions in the dynamics of a system. We estimated the inflection point directly from the raw data following the methodology of Christopoulos (2012). In the context of this paper, the inflection point corresponds to the transition from a convex strain-stress response to a concave one. All statistical and network analyses were performed with the R platform (R Core Team, 2012). The scripts are available from heading author upon request. This study was approved by the Ethics Committee of Universidad Nacional de Tucumán.

3 Results

The Fig. 2 shows the distributions of the fibrils according to their size. The box plots show a segregation of the data in three pools clearly distinguishable: one composed by small fibrils (the majority less than 50 nm), another conformed by medium fibrils (most of them between 50 and 100 nm), and other by large fibrils (the majority of them larger 100 nm). Segregation of the taxa proceeds regardless of the phylogeny. The range of variability of fibril size seems to increase in direct relationship with the median of fibrils.

3.1 Percolation

The Fig. 3 shows two scenes taken from the progressive recruitment process, namely (i) the scene concerning to the stage near to the formation of the percolating cluster, (ii) the scene concerning to the stage immediately posterior to the formation of the percolating cluster. The percolation corresponds to the inflection point in the fitted S-shaped curves relating quantile probabilities and fraction of spanning area. Percolation thresholds calculated for the empirical cases are in Table 1.

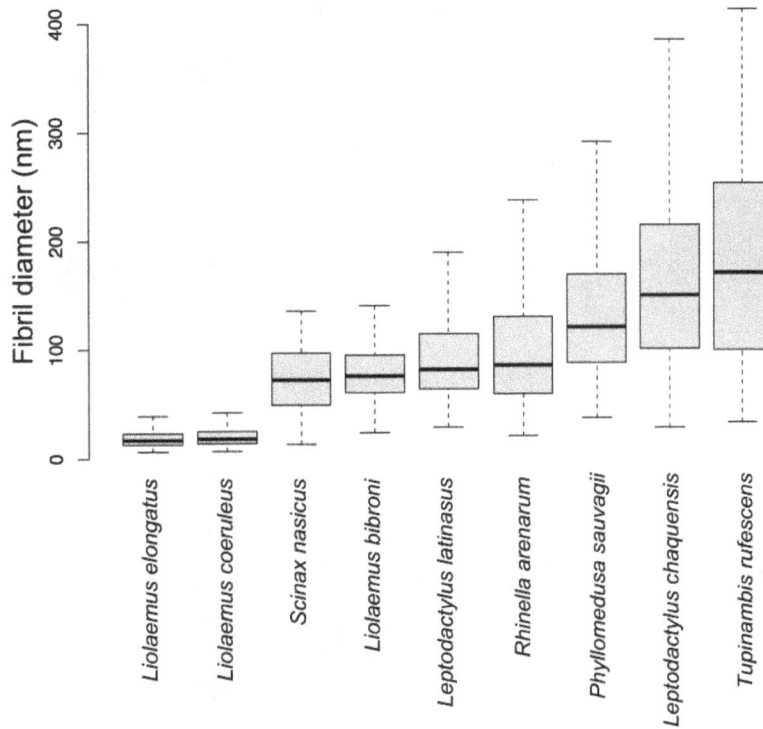

Fig. 2 Size distribution of digital flexor tendon fibrils across some representatives of South American herpetofauna. For each species, data coming from all samples were pooled into a single data set. Species can be segregated into three size categories: one dominated by small fibrils (< 50 nm), another characterized by intermediate fibrils (50-100 nm) and the last one where large fibrils prevail (> 100 nm). Outliers have been removed and whiskers extend to the most extreme data point which is no more than 1.5 times the interquartile range from the box.

Table 1 Percolation characteristics of tendon cross sections.

Species	Number of sampling images analyzed	Average percolation threshold		Percentage of images that resulted in a high percolation threshold (P< 0.05)	Median of the fibril size at the observed percolation threshold (nm)
		Trageted activation	Random activation		
Tupinambis rufescens	1	0.72	0.53	100%	241
Leptodactylus chaquensis	11	0.62	0.51	82%	187
Phyllomedusa sauvagii	7	0.65	0.48	100%	153
Rhinella arenarum	3	0.63	0.49	100%	92
Leptodactylus latinasus	5	0.59	0.49	80%	89
Scinax nasicus	1	0.65	0.48	100%	88
Liolaemus bibroni	1	0.59	0.47	100%	83
Liolaemus coeruleus	1	0.64	0.52	100%	36
Liolaemus elongatus	1	0.70	0.49	100%	23

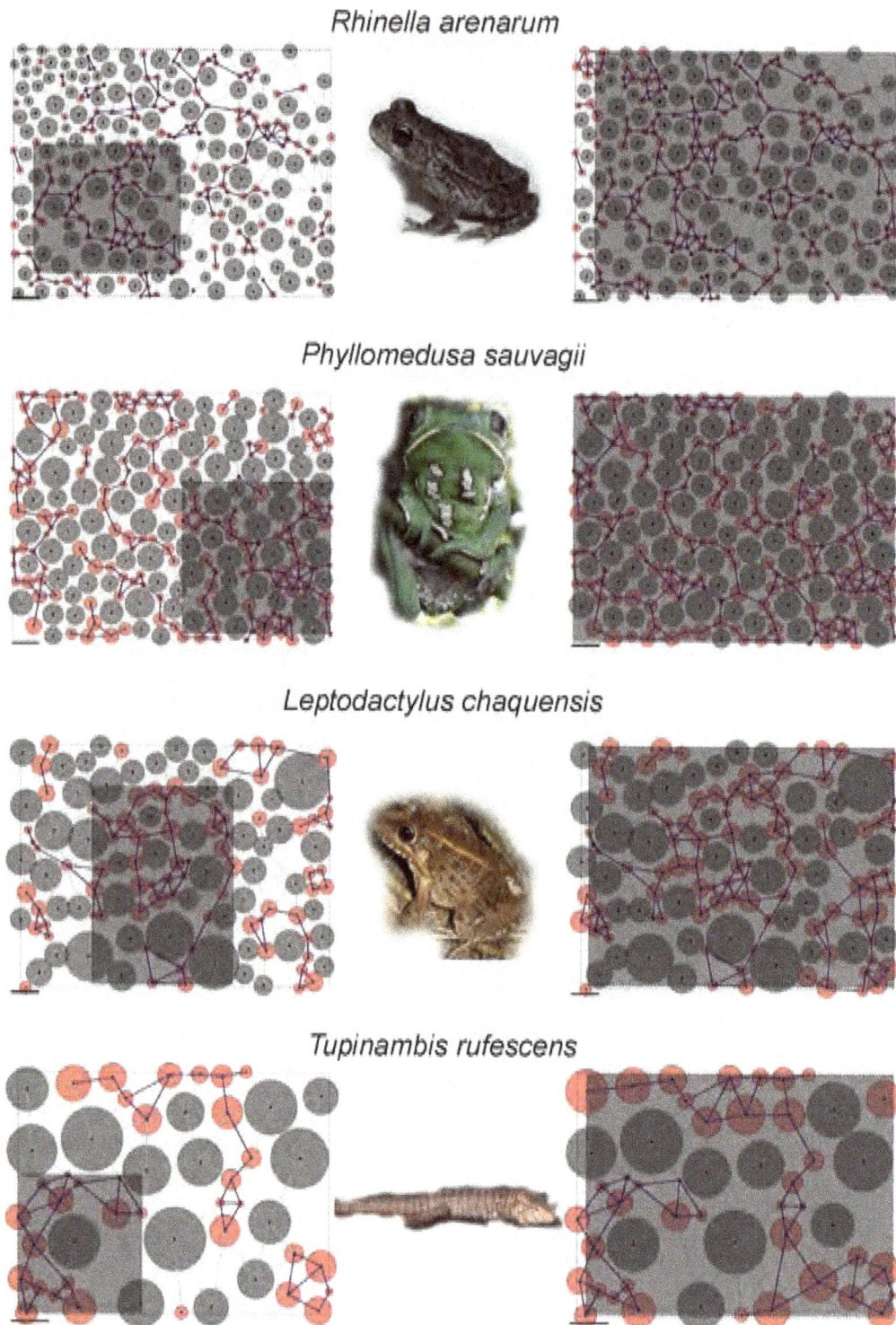

Rhinella arenarum

Phyllomedusa sauvagii

Leptodactylus chaquensis

Tupinambis rufescens

Fig. 3 Fibril proximity networks at two contrasting stages of node activation. Each fibril is proportionally represented to its cross section area. The underlying Gabriel graph (overall inferred network) is also shown and its nodes are located at the mass center of each fibril. Node activation proceeds orderly progressing from the smallest node to the largest one. Images at the left column represent snapshots of networks captured from activation instances close to the formation of the respective percolating clusters. Images at the right column illustrate scenes of fibril recruitment once the percolation threshold had been recently surpassed. As nodes are activated they shift their filling color from grey to red. There exists a blue link between a pair of adjacent nodes if both of them are activated; otherwise they are connected through a grey link. In each scene, dotted bounding boxes for both the study area and the largest component of activated nodes are displayed. Note that the area spanned by the bounding box of the putative percolating cluster (right column) closely matches the entire study area despite some nodes are still inactivated. Scale bars: 250 nm.

In the bulk of data of random simulation, the observed value of percolation thresholds was higher than random expectancy ($P < 0.05$) (Table 1).Random simulations revealed that there is a geometrical pattern linked to the size of the fibrils in most of the samples, because the percolation occurs at earlier stages of the progressive recruitment when the attribute size for the fibrils is randomly decoupled from the true location of them (Fig. 4). With a slight abuse of words, the high values of percolation thresholds detected imply a geometrical pattern constrained by size.

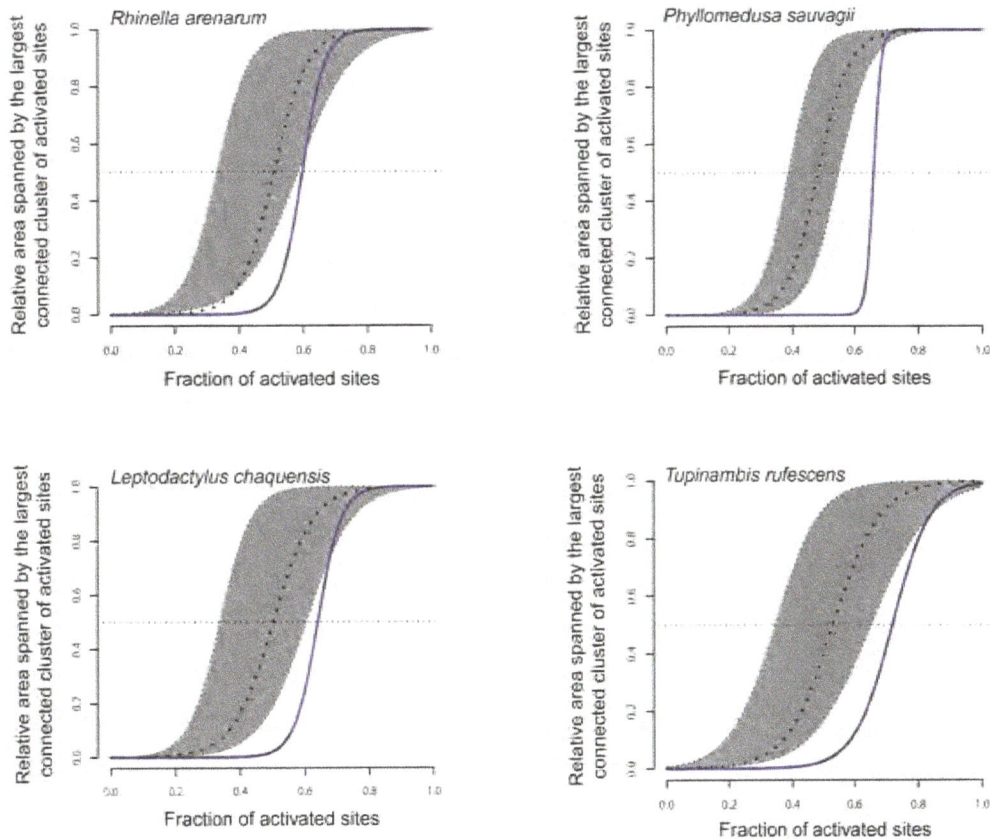

Fig. 4 Percolation curves. The percolation threshold is estimated as the inflection point of the adjusted S-shaped curve that relates the fraction of activated fibrils against the relative rectangular area occupied by the largest component of them. The dotted horizontal line passes through the inflection points of fitting curves. For the scenario of progressive size-dependent recruitment of fibrils (blue curve), the respective percolation threshold falls always beyond the one-sided 95% confidence interval. Confidence intervals were created after considering the curve parameters associated to 100 scenarios of random recruitment of fibrils.

3.2 Balanced interspersion of polymorphic fibrils

We have been able to reproduce conditions of delayed percolation through a non-random layout of heteromorphic units (elements of different sizes) on a square lattice. The pattern under consideration is called by us as Balanced Interspersion of Polymorphic Units (BIPU). BIPU consists of a regular alternated occurrence of larger and smaller fibrils throughout the physical dimensions of the system. Once the pool of fibrils has been partitioned into two sets of contrasting sizes (large and small), we allocate them following a chess board pattern (Fig. 5). Sequential activation of sites in this pre-ordered template may achieve a percolation threshold similar to the observed ones in the analysis of cross-section images of tendon ultrastructure, i.e. $q \sim 0.64$ (Table 2).

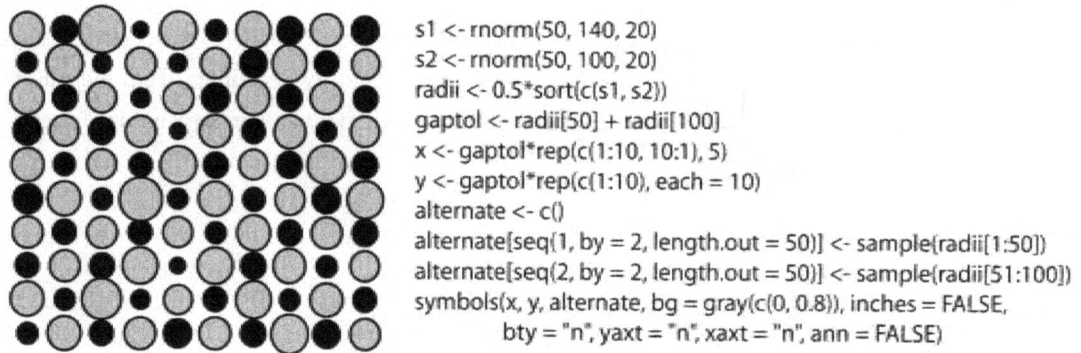

```
s1 <- rnorm(50, 140, 20)
s2 <- rnorm(50, 100, 20)
radii <- 0.5*sort(c(s1, s2))
gaptol <- radii[50] + radii[100]
x <- gaptol*rep(c(1:10, 10:1), 5)
y <- gaptol*rep(c(1:10), each = 10)
alternate <- c()
alternate[seq(1, by = 2, length.out = 50)] <- sample(radii[1:50])
alternate[seq(2, by = 2, length.out = 50)] <- sample(radii[51:100])
symbols(x, y, alternate, bg = gray(c(0, 0.8)), inches = FALSE,
        bty = "n", yaxt = "n", xaxt = "n", ann = FALSE)
```

Fig. 5 Model of balanced interspersion of polymorphic fibrils. One hundred fibrils have been sampled from two different distributions of cross section diameter in nanometers. Fifty elements were drawn from a normal variable $S_1 \sim (140, 20)$ whereas the other fifty elements were sampled from $S_2 \sim (100, 20)$. Subsequently, the totality of elements is arranged across a 10 X 10 square lattice where the larger elements (in any order) alternate with the smaller ones (in any order) in a chess board pattern. For an easy backtracking of ideas, the generating R script is displayed next to the respective layout.

Table 2 Statistical synthesis (mean plus minus standard deviation) of percolation experiments performed on several hypothetical square lattices in which putative fibrils of various sizes were allocated following the pattern of balanced interspersion (chess board layout where larger fibrils alternate side-to-side with smaller fibrils). For each lattice configuration, 100 replications were run.

Size of the square lattice	Targeted activation	Random activation
10 X 10	0.65 (± 0.03)	0.52 (± 0.05)
10 X 11	0.66 (± 0.03)	0.52 (± 0.05)
10 X 12	0.65 (± 0.03)	0.51 (± 0.05)
11 X 11	0.66 (± 0.03)	0.51 (± 0.05)
11 X 12	0.66 (± 0.03)	0.52 (± 0.05)
11 X 13	0.67 (± 0.03)	0.52 (± 0.05)
12 X 12	0.66 (± 0.03)	0.52 (± 0.06)
12 X 13	0.65 (± 0.03)	0.52 (± 0.05)
12 X 14	0.66 (± 0.03)	0.53 (± 0.05)
13 X 13	0.66 (± 0.03)	0.52 (± 0.05)
13 X 14	0.66 (± 0.03)	0.52 (± 0.05)
13 X 15	0.66 (± 0.03)	0.52 (± 0.04)
14 X 14	0.66 (± 0.03)	0.53 (± 0.04)
14 X 15	0.67 (± 0.03)	0.52 (± 0.05)
14 X 16	0.66 (± 0.02)	0.52 (± 0.05)
15 X 15	0.66 (± 0.03)	0.53 (± 0.05)
15 X 16	0.66 (± 0.02)	0.52 (± 0.04)
15 X 17	0.66 (± 0.02)	0.53 (± 0.04)
16 X 16	0.66 (± 0.03)	0.53 (± 0.05)
16 X 17	0.66 (± 0.03)	0.53 (± 0.04)
16 X 18	0.66 (± 0.02)	0.53 (± 0.04)
17 X 17	0.67 (± 0.03)	0.53 (± 0.04)
17 X 18	0.66 (± 0.02)	0.53 (± 0.04)

17 X 19	0.66 (± 0.02)	0.53 (± 0.03)
18 X 18	0.66 (± 0.02)	0.53 (± 0.04)
18 X 19	0.67 (± 0.02)	0.53 (± 0.04)
18 X 20	0.67 (± 0.02)	0.54 (± 0.04)

3.3 Stress-strain relationship and inflection point

Typical stress-strain curves for anurans and squamatans digital flexor tendons are shown in Fig.6. Ultimate tensile stress (UTS) for each species and summary statistics about the stress and strain found at the inflection point are showed in Table 3.

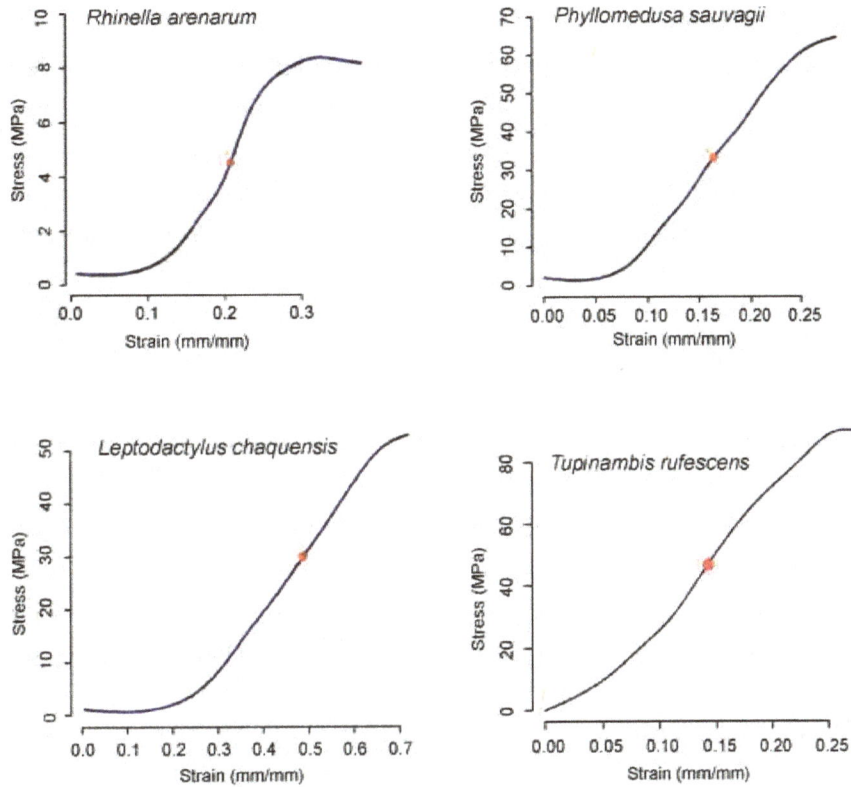

Fig. 6 Stress-strain curves. Direct readings from the Instron® device were adjusted through piecewise-cubic splines. Inflection points marked as red dots.

Table 3 Basic measurements of stress-strain relationship. Reported values are the medians of the respective samples. Scores of critical strain and stress are those recorded at the inflection point of the polynomial cubic fit of the empirical data. Maximum stiffness corresponds to the slope of the tangent at such inflection point. UTS= ultimate tensile strength.

Species	Strain at UTS (mm/mm)	UTS (MPa)	Criticalstrain (mm/mm)	Critical stress (MPa)	Maximumstiffness (MPa)
Rhinella arenarum (n = 6)	0.36	11.22	0.25	6.64	81.27
Phyllomedusa sauvagii (n = 5)	0.23	65.50	0.16	50.14	569.65
Leptodactylus chaquensis (n = 6)	0.62	51.49	0.45	29.82	176.56
Tupinambis rufescens (n = 5)	0.27	47.52	0.15	21.89	488.13

4 Discussion

The survival of living organisms is dependent on the functional integrity of all their organs and organ systems. The tendon is a biomechanical system of force transference from muscles to bones in which its functional integrity translates into a controlled movement of joints. Without loss of generality, the functional integrity of any system relies on the ability of their components to propagate information and offer an adaptive response to external influences. It is hard to conceive such a property in a system of poorly connected components or, equivalently stated, in a system with a relaxed network structure underpinning it. In this paper, we have assumed a network organization for the fibrils of the tendon and consequently explored the patterns of connections between them under the approach of network percolation. At least to our knowledge, percolation theory is the most adequate conceptual framework to answer inquiries about the functional integrity of the tendon, because it deals explicitly with the subject of information propagation throughout the physical dimensions where the system resides.

While Svensson et al. (2013) have recently pointed out that fibrils are not evenly loaded within the tendon butare sequentially recruited throughout the initial stress-strain region, the analysis of our data revealed moreover a size-dependent effect for that sequential recruitment of fibrils. The percolation thresholds are consistently biased towards the upper tail of the statistical distribution of the size of fibrils. It is then comprehensible that the following taxa: *Leptodactylus chaquensis*, *Phyllomedusa sauvagii*, and *Tupinambis rufescens*, exhibit great values for the percolation threshold size. All they surpass easily the 100 nm of diameter in the cross section of fibrils. Ultimately, this means that taxa with great fibrils do not achieve percolation by activation of their small fibrils alone, larger fibrils are also necessary to be activated. A marginal essay performed by us using images about tendon ultrastructure available from literature (horse: Parry, 1988; mouse: Ameye et al., 2002; rabbit: Gill et al., 2004) showed the same pattern. In mammals it also seems to be necessary the activation of fibrils around the 65th percentile to achieve percolation. When compared with the random activation of fibrils, the observed values for percolation thresholds also resulted significantly higher than random expectations.

The delay detected for the occurrence of percolation during the recruitment process indicates that collagen fibrils are spatially arranged according to a geometrical non-random model, in which the location and size attribute are both important. We propose the model BIPU of fibril arrangement characterized by the even interspersion of fibrils of different size category throughout the ECM. This pattern would account for a uniform fibrillar density, that makes restricted regions to be very similar with regards of the overall tendon, and also would explain the delayed percolation threshold observed along our experiments of targeted sites activation. In fact, the balanced interspersion of polymorphic unit simply that larger fibrils makes a shadow effect on nearby smaller fibrils decreasing thus the chance of direct connections between the latter ones. Additionally, the balanced interspersion would also imply that heterogeneous fibrilsare close each other facilitating the access to information managed by the different fibrils. An open question is if the delay for percolation represents a way to optimize the physiological range of tendons.

Traditionally, a typical stress-strain curve for a tendon has been characterized by three regions: the toe, linear and non-linear regions (Wang, 2006). On the contrary, we interpret the resulting stress-strain curve as composed by two phases: the convex and concave regions, being the point of inflection the transition between them. Functionally, they would reflect two quite different behaviors: in the first region, stiffness increases by a targeted activations of fibrils depending upon their sizes, whereas in the second region the tendon begins to yield and fracture. This change in the approach to analyze the stress-strain curve abruptly breaks with the conventional perspective of considering an intermediate linear region in which the respective slope reflect some biomechanical property intrinsically linked to the structure of tendons. In other words, we think that the task of calculating the Young's modulus as a unique and distinctive measure seems to be a sum-zero exercise in analyzing functional tendon properties. Consequently, we propose to rely on the point of inflection to calculate several metrics that account for the mechanical properties of tendons such as critical stress, critical strain (physiological range), and critical stiffness. We propose an isomorphic linkage between stress-strain and percolation curves, and we hypothesize that the inflection points of both curves reflect a phase transition associated to the same underlying process. This process is probably the propagation of fibril disfunction throughout the extracellular ground of connective tissue. When failure percolates the overall system begins to elongate elastically, so the percolation threshold could indicate the upper bound for the physiological range of the tendon.

In our approach, the physical properties of a tendon depend on the pattern of connections behind the geometrical network of fibrils where location and difference in size play a key role. It is interesting to note that a polimodal distribution of fibril sizes characterize those connective tissues able to resist tensional forces such as tendons. On the contrary, those connective tissues that are commonly under no tensile force show a typically unimodal distribution of fibril size (e.g. buccal gingival mucosa collagen: Ottani et al., 1998; skin collagen: Danielson et al., 1997; Silver et al., 2001; corneal stroma lamella: Parry et al., 1978). Likewise, those tendons that are not yet functional, such as embryonic tendons exhibit also anunimodal distribution of collagen fibril size (Fleischmajer et al., 1988; Parry et al., 1978; Zhang et al., 2005). Unimodal distribution of collagen fibril size is also present in regenerated tendons (Gill et al., 2004). We suggest that for acting efficiently, the polymorphism must be accompanied of another geometrical feature of spatial organization. This spatial organization seems to be subsumed into a pattern of balanced interspersion of different fibrils throughout the ECM. An immediate application of our proposal concerns with the design of strategies for tissue engineering provided of biomimetic-synthetic nanofibrous composites.

5 Concluding Remarks

The polymorphic nature of collagen fibrils in addition to the presence of molecular cross-links between them lead us to think in an heterogeneous network of fibrils that influences the mechanical behavior of tendon as a whole. The main contribution of our work is to combine geometrical and network considerations into a single framework by using the percolation approach. This analysis allows a holistic study of the structural properties of a tendon based on their architectural design. The geometrical pattern we have suggested (non-random packing constrained by size) is amenable with the idea of a progressive and sequential recruitment of fibrils dependent on their size. This pattern offers a delay to percolation, which could act a mechanism for expanding the physiological range. As a surplus of our work, we also address the circumstance of being percolation curves mirrored by the empirical curves of stress-strain obtained from the same studied tendons. By virtue of this isomorphism, we hypothesize that the inflection points of both curves are different quantitative manifestations of a common transitional process during mechanical load transference.

Appendix Specimens examined

L: personal collection of María Laura Ponssa; FBC: personal collection of Felix B. Cruz; FML: Fundación Miguel Lillo; GS: personal collection of Gustavo Scrocchi; ST: personal collection of Sebastián Torres. An asterisk is appended to the access number if the respective specimen was used for studying cross sectional tendon images, otherwise specimens were used for stress-strain analysis.
Leptodactylus chaquensis: L103, L766, L850*, L344, L950, L950-951*, L964*, L971*, ST103; *Leptodactylus latinasus*: L937a*, L939-942*, L944-945*;
Phyllomedusa sauvagii: L849*, L936*, L938*, L946-947, L947(1), L946-948*, L971;

Rhinella arenarum: L51, L851*, L909, L935, L935*, L937b*, L961, L965;

Scinax nasicus: L949*;

Liolaemus bibroni: FBC 1265*;

Liolaemus coeruleus: FBC 1265*;

Liolaemus elongatus: GS3227;

Tupinambis rufescens: FML 07256*, FML7554.

Acknowledgements

This work was supported by the ANPCyT and CONICET through the following research grants: PICT 2012-1067, PICT 2012-1910, PIP 112-200801-00225, and BID-PICT 606. All authors are grateful to CONICET for supporting our work via its program of post-graduate fellowships and research grants. We thank Gabriela Pacios (Facultad de Odontología, Universidad Nacional de Tucumán)for assistance with Instron, and NicolásNieva (Facultad de Ciencias Exactas, Universidad Nacional de Tucumán) for comments that greatly improved the structure of the manuscript.

References

Ameye L, Aria D, Jepsen K, et al. 2002. Abnormal collagen fibrils in tendons of biglycan/fibromodulin-deficient mice lead to gait impairment, ectopic ossification, and osteoarthritis. FASEB J, 16: 673-680

Battaglia TC, Clark RT, Chhabra A, et al. 2003. Ultrastructural determinants of murine achilles tendon strength during healing. Connective Tissue Research, 44: 218-224

Berthod F, Germain L, Li H, et al. 2001. Collagen fibril network and elastic system remodeling in a reconstructed skin transplanted on nude mice. Matrix Biology, 20: 463-473

Berkowitz B, Ewing RP. 1998. Percolation theory and network modeling applications in soil physics. SurvGeophys, 19: 23-72

Callaway DS, Newman MEJ, Strogatz SH, Watts DJ. 2000. Network robustness and fragility: percolation on random graphs. Physical Review Letters, 85: 5468-5471

Chandran PL, Barocas VH. 2006. Affine versus non-affine fibril kinematics in collagen networks: theoretical studies of network behavior. Journal of Biomechanical Engineering, 128: 259-270

Christopoulos DT. 2012. Developing methods for identifying the inflection point of aconvex/ concave curve. arXiv:1206.5478v1 [math.NA]

Cuestas E, Vilaró M, Serra P. 2011.Predictibilidad de la propagación espacial y temporal de la epidemia de influeza A-H1N1 en Argentina por el método de percolación. Revista Argentina de Microbiología, 43: 186-190

Danielson KG, Baribault H, Holmes DF, et al. 1997. Targeted disruption of decorin leads to abnormal collagen fibril morphology and skin fragility. Journal of Cell Biology, 136: 729-743

Ferrarini A. 2013. Exogenous control of biological and ecological systems through evolutionary modeling. Proceedings of the International Academy of Ecology and Environmental Sciences, 3(3): 257-265

Ferrarini A. 2014. True-to-life friction values in connectivity ecology: Introducing reverse flow connectivity. Environmental Skeptics and Critics, 3(1): 17-23

Fleischmajer R, Perlish JS, Timpl R, et al. 1988. Procollagen intermediates during tendon fibrillogenesis. Journal Hist Cyt, 36: 1425-1432

Gabriel KR, Sokal RR. *1969*. A new statistical approach to geographic variation analysis. Systematic Zoology,

18:259-278

Gill SS, Turner MA, Battaglia TC, et al. 2004. Semitendinosus regrowth. Biochemical, ultrastructural, and physiological characterization of the regenerate tendon. American Journal of Sports Medicine, 32: 1173-1181

Grimmett G. 1999. Percolation (2nd edition). Springer-Verlag, New York, USA

Jaromczyk JW, Toussaint GT. 1992. Relative neighborhood graphs and their relatives. Proceedings of the IEEE, 80: 1502-1517

Julkunen P, Livarinen J, Brama PA, et al. 2010. Maturation of collagen fibril network structure in tibial and femoral cartilage of rabbits. Osteoarthritis and Cartilage, 18: 406-415

Långsjö TK, Arita M, Helminen HJ. 2009. Cartilage collagen fibril network in newborn transgenic mice analyzed by electron microscopic stereology. Cells, Tissues and Organs, 190: 209-218

Långsjö TK, Vasara AI, Hyttinen MM, et al. 2010. Quantitative analysis of collagen network structure and fibril dimensions in cartilage repair with autologous chondrocyte transplantation. Cells, Tissues and Organs, 192: 351-360

Mikic B, Schalet BJ, Clark RT, et al. 2001. GDF-5 deficiency in mice alters the ultrastructure, mechanical properties and composition of the Achilles tendon. Journal of Orthopaedic Research, 19: 365-371

Newman MEJ. 2010. Networks: An Introduction. Oxford University Press, Oxford, UK

Newman MEJ, Ziff RM. 2001. A fast Monte Carlo algorithm for site or bond percolation. Physical Review E, 64: 016706

Ottani V, Franchi M, Depasquale V, et al. 1998. Collagen fibril arrangement and size distribution in monkey oral mucosa. Journal of Anatomy, 192: 321-328

Parry DAD. 1988. The molecular and fibrillar structure of collagen and its relationship to the mechanical properties of connective tissue. Biophysical Chemistry, 29: 195-209

Parry DAD, Barnes GRG, Craig AS. 1978. A comparison of the size distribution of collagen fibrils in connective tissues as a function of age and a possible relation between fibril size distribution and mechanical properties. Proceedings of the Royal Society of London B, 203: 305-321

Pawłowska M, Sikorski A. 2013.Monte Carlo study of the percolation in two-dimensional polymer systems. Journal of Molecular Modeling (DOI 10.1007/s00894-013-1892-y).

Purslow PP, Wess TJ, Hukins DWL. 1998. Collagen orientation and molecular spacing during creep and stress-relaxation in soft connective tissues. Journal of Experimental Biology, 201: 235-242

R Core Team. 2012. R: A language and enviroment for statistical computing. R Foundation for statistical computing Vienna, Austria. Available at: http://www.R-project.org/.

Reed CC, Iozzo RV. 2002. The role of decorin in collagen fibrillogenesis and skin homeostasis. Glycoconjugate Journal, 19: 249-255

Rigozzi S, Müller R, Snedeker JG. 2010. Collagen fibril morphology and mechanical properties of the *Achilles tendon* in two inbred mouse strains. Journal of Anatomy, 216: 724-731

Rigozzi S, Stemmer A, Müller R, Snedeker JG. 2011. Mechanical response of individual collagen fibrils in loaded tendon as measured by atomic force microscopy. Journal of Structural Biology, 176: 9-15

Shannon CE. 1948. A mathematical theory of communication. Bell System Technical Journal, 27: 379-423, 623–656.

Shirazi R, Vena P, Sah RL, Klisch SM. 2011. Modeling the collagen fibril network of biological tissues as a nonlinearly elastic material using a continuous volume fraction distribution function. Mathematics and Mechanics of Solids, 16: 707-716

Silver FH, Freeman JW, Devore D. 2001. Viscoelastic properties of human skin and processed dermis. Skin

Research and Technology, 7: 18-23

Svensson RB, Mulder H, Kovanen V, et al. 2013. Fracture mechanics of collagen fibrils: influence of natural cross-links. Biophysical Journal, 104: 2476-2484

Teixeira-Filho PF, Rocha-Barbosa O, Paes V, et al. 2001. Ecomorphological relationships in six lizard species of Restinga da Barra de Marica, Rio deJaneiro, Brazil. Rev Chil Anat, 19: 45-50

Tulli MJ, Abdala V, Cruz FB. 2011.   Relationships among morphology, clinging performance and habitat use in Liolaemini lizards. Journal of Evolutionary Biology, 24: 843-855

Wainwright SA, Biggs WD, Currey JD, Gosline JM. 1976. Mechanical Design in Organisms. Edward Arnolds, London, UK

Wang JHC. 2006. Mechanobiology of tendon. Journal of Biomechanics, 39: 1563-1582

Zhang G. 2005. Evaluating the viscoelastic properties of biological tissues in a new way. Journal of Musculoskeletal and Neuronal Interactions, 5: 85-90

Zhang G, Young BB, Ezura Y, et al. 2005. Development of tendon structure and function: regulation of collagen fibrillogenesis. Journal of Musculoskeletal and Neuronal Interactions, 5: 5-21

Zhang WJ. 2012a. Computational Ecology: Graphs, Networks and Agent-based Modeling. World Scientific, Singapore

Zhang WJ. 2012b. Modeling community succession and assembly: A novel method for network evolution. Network Biology, 2(2): 69-78

Zhang WJ. 2012c. Several mathematical methods for identifying crucial nodes in networks. Network Biology, 2(4): 121-126

Zhang WJ. 2013. Network Biology: Theories, Methods and Applications. Nova Science Publishers, New York, USA

7

Functional interactome of Aquaporin 1 sub-family reveals new physiological functions in *Arabidopsis Thaliana*

Mohamed Ragab Abdel Gawwad[1], **Jasmin Šutković**[1], **Lavinija Mataković**[2], **Mohamed Musrati**[1], **Lizhi Zhang**[3]

[1]Genetics and Bioengineering department, International University of Sarajevo, Ilidza, 71220 Bosnia and Herzegovina

[2]University of Josip Jurja Strossmayer, Biology Department, Croatia

[3]Department of Molecular Genetics, The Ohio State University, 484 West 12th Avenue, Columbus, OH 43210, USA

E-mail: mragab@ius.edu.ba

Abstract

Aquaporins are channel proteins found in plasma membranes and intercellular membranes of different cellular compartments, facilitate the water flux, solutes and gases across the cellular plasma membranes. The present study highlights the sub-family plasma membrane intrinsic protein (PIP) predicting the 3-D structure and analyzing the functional interactome of it homologs. PIP1 homologs integrate with many proteins with different plant physiological roles in *Arabidopsis thaliana* including; PIP1A and PIP1B: facilitate the transport of water, diffusion of amino acids and/or peptides from the vacuolar compartment to the cytoplasm, play a role in the control of cell turgor and cell expansion and involved in root water uptake respectively. In addition we found that PIP1B plays a defensive role against *Pseudomonas syringae* infection through the interaction with the plasma membrane Rps2 protein. Another substantial function of PIP1C via the interaction with PIP2E is the response to nematode infection. Generally, PIP1 sub-family interactome controlling many physiological processes in plant cell like; osmoregulation in plants under high osmotic stress such as under a high salt, response to nematode, facilitate the transport of water across cell membrane and regulation of floral initiation in *Arabidopsis thaliana*.

Keywords Aquaporins; *Arabidopsis thaliana*; interactome; 3-D structure.

1 Introduction

Transport of materials across biological membranes is a fundamental process in all living cells. Charged and polar molecules, however, require special pathways to cross the cellular membrane, as the hydrophobic tails of lipid molecules create a considerable energetic barrier against their diffusion. Membrane channels provide such pathways for selective exchange of water-soluble materials (water, ions, and other nutrients) across the membrane. Three major functional characteristics of membrane channels, which are furnished by specific arrangements of amino acids in their structure, are permeation, selectivity, and gating. Aquaporins (AQP) are a family of membrane channels primarily responsible for conducting water across cellular membranes. These channels are widely distributed in all kingdoms of life, including bacteria, plants, and mammals. They form tetramers in the cell membrane, and facilitate the transport of water and, in some cases, other small solutes across the membrane (Wang and Tajkhorshid, 2007; Li et al., 2011). These proteins are members of the larger family of major intrinsic proteins (MIPs) with 26–34 kDa, and all protein members have six transmembrane helices, with both N- and C-termini on the cytosolic side of the membrane (Fraysse et al., 2005).

Water permeation through aquaporins is a passive process that follows the direction of osmotic pressure across the membrane. Although many aquaporins function as always-open channels, a subgroup of aquaporins, particularly in plants have evolved a sophisticated molecular mechanism through which the channel can be closed in response to harsh conditions of the environment, under which exchange of water can be harmful for the organism (Törnroth-Horsefield et al., 2005). Plants respond to drought or flood conditions by shutting down almost all of their AQP .In humans, 13 different AQP (AQP0–AQP12) have been characterized in various organs (kidneys, eyes, and the brain). The solved structures of several AQP at high resolution are indicative of a conserved protein architecture in the whole family (Wang and Tajkhorshid, 2007).

Tetramerization is a common structural feature of AQPs (Yu et al., 2006). Four functionally independent pores provide highly selective pathways for water permeation across the low dielectric barrier of lipid bilayers. Another pore, known as the central pore, is formed between the 4 monomers (Muller et al., 2002).

Plant aquaporins are categorized as either tonoplast intrinsic proteins (TIPs) or plasma membrane intrinsic proteins, PIPs (Chaumont et al., 2000). Among the members of the plant MIP family, the PIPs form the most highly conserved subfamily (Fraysse et al., 2005). The PIP subfamily can be further subdivided into two groups PIP1 and PIP2, of which the PIP1 isoforms are most tightly conserved, sharing >90% amino acid sequence identity (Fraysse et al., 2005).

The members of these groups differ in N- and C-termini lengths, the N-terminus being longer in PIP1 aquaporins. In plant cells, the transport of PIP proteins to the plasma membranes or the integration in the membrane appears to be dependent on PIP2 expression (Zelazny et al., 2007). The genome of *Arabidopsis thaliana* encodes 35 full-length aquaporin homologues. Thirteen of them belong to the PIP subfamily (Santoni et al., 2003).

The importance of aquaporins in environmental stress responses has been demonstrated through gene expression analyses of various plants species and the characterization of transgenic plants expressing these aquaporins.

Aquaporin-1 (AQP1), a membrane channel protein, is the first characterized member of the aquaporin (AQP) family (Hohmann et al., 2000). The protein is abundantly present in multiple human tissues, such as the kidneys. AQP1 forms homotetramers in cell membranes, each monomer forming a functionally independent water pore, which does not conduct protons, ions, or other charged solutes.

Members of the AQP1 family include: PIP 1-1, PIP1-2, PIP1-3, PIP1-4 and PIP1-5. The PIP1 subfamily of aquaporins constitutes about 1% of the plasma membrane (PM) proteins from Arabidopsis thaliana leaves (Robinson et al., 1996).

The members of this family are involved in various biological processes including Golgi organization, calcium ion transport, glycolysis, hyperosmotic response and in response to cadmium ion. Furthermore, it is shown that these proteins have a function in response to fructose and temperature stimulus. These proteins are expressed ubiquitously and during leaf development the protein level decreases slightly and their function are impaired by Hg (2+) (Chaumont et al., 2000; Hohmann et al., 2000; Zelazny et al., 2007).

Biologically PIP1 subfamily proteins respond to salt stress. It was shown that these proteins are also involved in other processes, such as acetyl-CoA metabolic process, brassinosteroid biosynthetic process, sterol biosynthetic process and in cellular response to iron ion starvation and iron ion transport. Furthermore, they have important role in carbon dioxide transport and regulation of protein localization (Hohmann et al., 2000; Li et al., 2010).

The subfamily of PIP1 proteins is located in chloroplast envelope, integral to membrane, membrane, mitochondria, plasma membrane, plasmodesma and vacuole. Analyses of mRNAs show that the two spinach (*Spinacia oleracea*), PIP1 isoforms, SoPIP1; 1 and SoPIP1; 2 according to the nomenclature proposed by Johanson, are differentially expressed (Johanson et al., 2001). Recent study reported that PIP1; 2, the most abundant PIP in spinach leaves, is localized in phloem sieve elements in source and sink tissues (Fraysse et al., 2005).

In this study, the 3-D structure of PIP1 homologs had been predicted in order to understand the functional interactome of its homologs. Moreover, new physiological functions had been assigned to Aquaporins generally and specifically PIP1protein sub-family.

2 Materials and Methods

2.1 Retrieval of PIP1 homologs protein sequences

PIP1 protein sequences were obtained from the sequence database of National center of Biotechnology information (NCBI) (Sayers et al., 2009) and the Arabidopsis Information Resource (TAIR) (Lamesch et al., 2012), as shown in Table 1.

2.2 Multiple sequence alignment and phylogenetic tree

Amino acid multiple sequence alignment was made in using ClustalW2 program (Larkin et al., 2007), the same program was used for phylogenetic tree construction.

2.3 3-D structure prediction and conformation

The aquaporin protein sequences of PIP1 and PIP2 subfamilies from *Arabidopsis thaliana* was obtained from NCBI. The FASTA sequence of Aquaporin proteins from Arabidopsis thaliana was obtained from NCBI and 3-D structure prediction was made for each PIP protein by Swiss Model server (Schwede et al., 2003) and then the models were visualized by PyMOL molecular visualization program (The PyMOL Molecular Graphics System).

Table 1 PIP1s subfamily members and their gene ID
codes from NCBI and TAIR.

PIP1 proteins	NCBI Gene IDs	TAIR IDs
PIP1-1	332646681	AT3G61430
PIP1-2	330255530	AT2G45960
PIP1-3	332189192	AT1G01620
PIP1-4	30023776	AT4G00430
PIP1-5	332659348	AT4G23400

2.4 Protein-protein interaction prediction

Protein interaction networks were determined by using STRING - Known and Predicted Protein-Protein Interactions (Franceschini and Szklarczyk, 2013). The functional interactome of PIP1 homologs was calculated at 0.7 confident factor.

2.5 Subcellular localization

Subcellular localization for each protein was predicted in Protein Localization Prediction Software (WOLF PSORT) (Letunic et al., 2012).

3 Results

3.1 Multiple sequence alignment and phylogenetic tree

The protein sequences of PIP1 aquaporin sub-family of *Arabidopsis thaliana* was obtained from the TAIR. Multiple alignment of the primary structure of the target proteins highlights the degree of sequence conservation and high sequence similarity. Moreover, conserved Asn-Pro-Ala (NPA) motifs were found in all PIP1 homologs (Fig. 1).

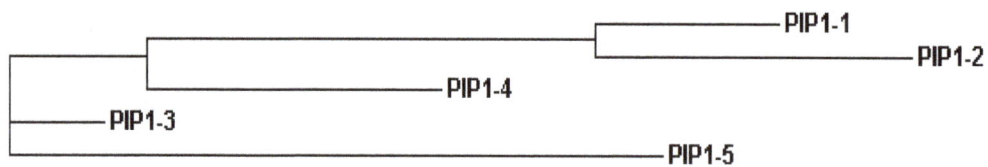

Fig. 1 Phylogenetic tree constructed by ClustalW2.

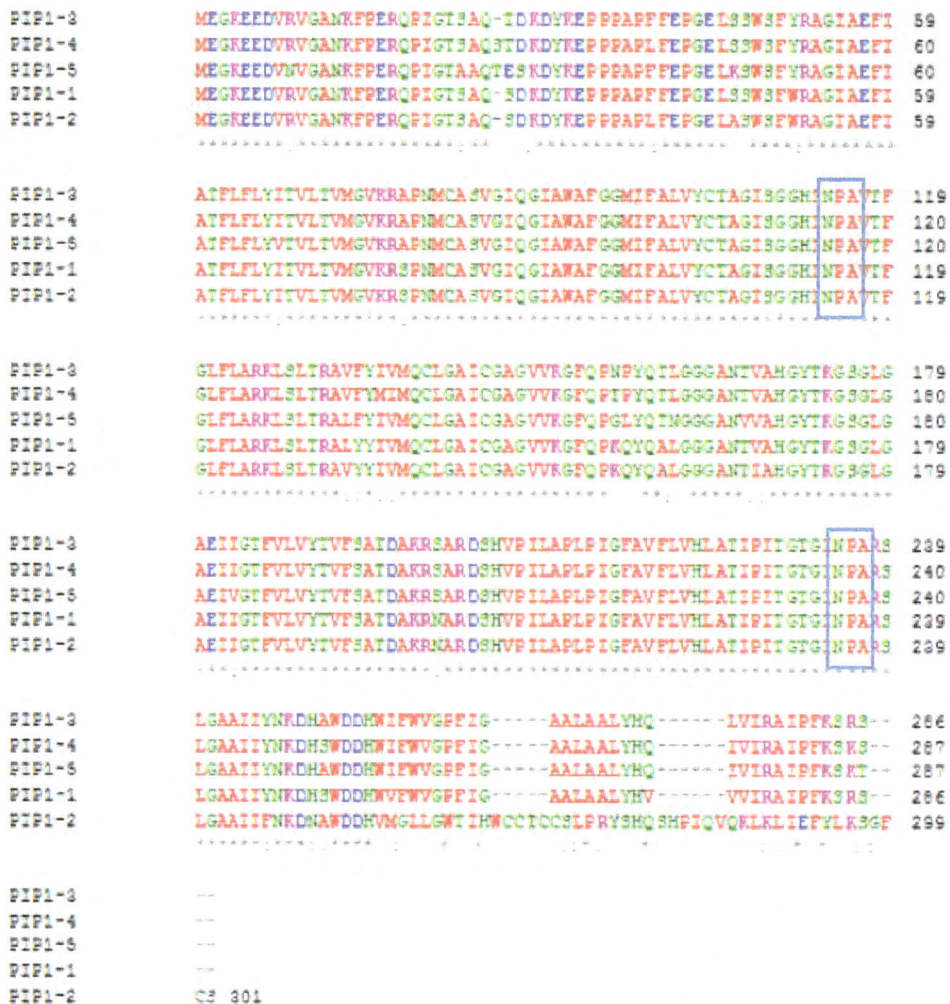

Fig. 2 Alignments of PIP1s by ClustalW2 software showing the common conserved NPA motifs in the boxs.

3.2 3-D structure prediction and conformation

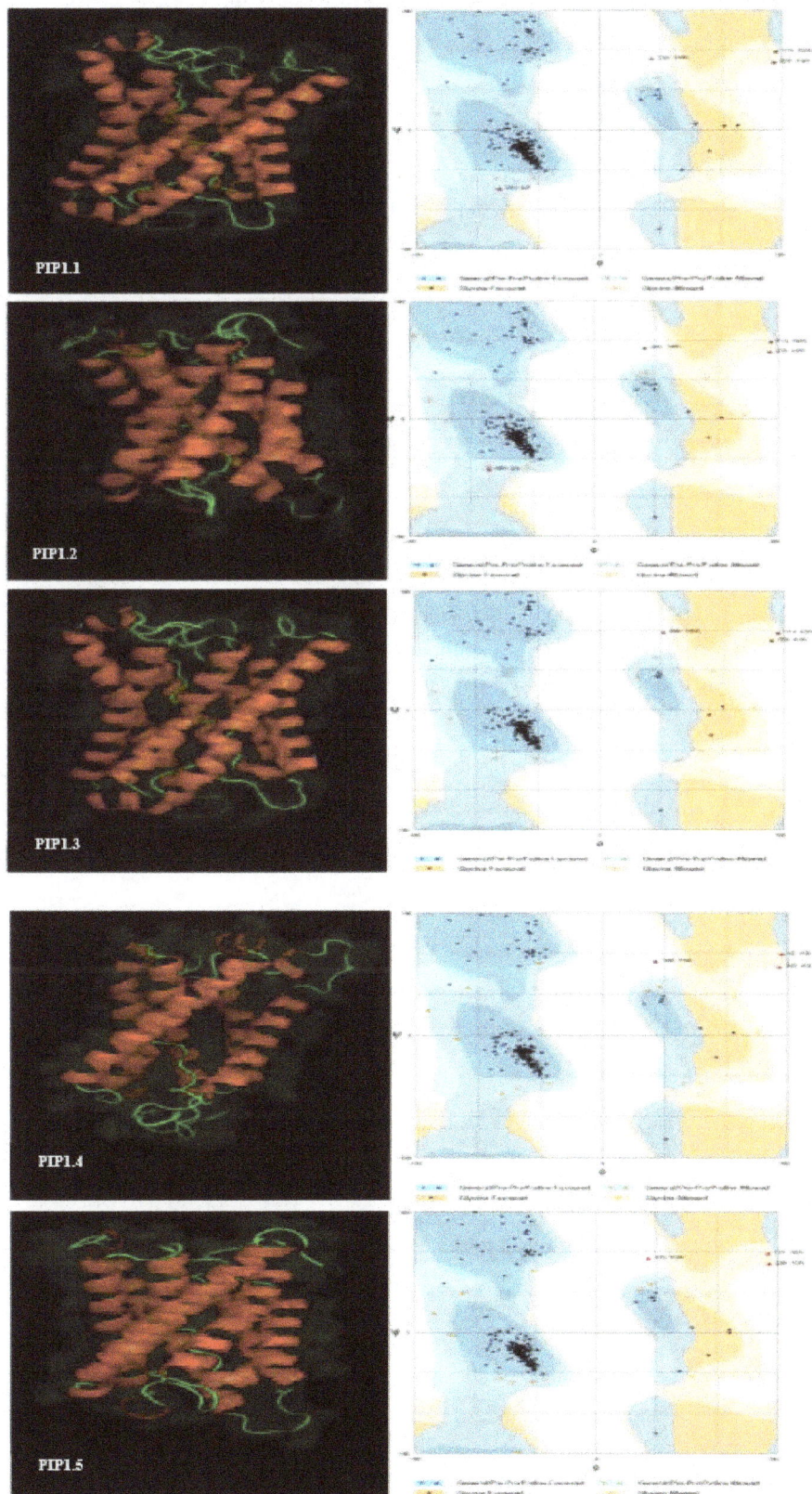

Fig. 3 3-D structure predictions of PIP1 subfamily homologs and related Ramachandran plots.

3.3 Subcellular localization

The Subcellular localizations of PIP1s are analyzed with using WoLF PSORT program. Results showed good and significant reliabilities that all of the subfamily; PIP1-1, PIP1-2, PIP1-3, PIP1-4, PIP1-5 are mainly located in the plasma membrane, alongside other sites including; cytoplsm, plastid, mitochondrion, golgi and E.R Table 2.

Table 2 Subcellular sites of PIP1s subfamily obtained from WOLF PSORT.

Protein	Subcellular localization
PIP1-1	Plasma membrane
PIP1-2	Plasma membrane, plastid
PIP1-3	Plasma membrane, plastid
PIP1-4	plasma membrane,mitochondrion
PIP1-5	Plasma membrane, gologi, E.R.

3.4 Interactome analysis

According to the data obtained from Simple molecular architecture tool, we found many interactions belonging to the five PIP1 protein sub-family in various cellular localizations, but mainly the partners were localized in the plasma membrane, the interactome network of PIP1 subfamily showed significant interaction with almost all members in PIP2 subfamily and PIP3 subfamily. Additionally strong interaction of PIP1A, PIP1C and PIP1; 5 is shown with RD28, DELTA-TIP, and ANAC098 whereas PIP1B weakly interacts with WOL and Rps2 and Aquaporin 3 family. PIP1; 4 showed significant interaction with RD28, PIP3 family and PIP2; 8 protein (Fig. 4).

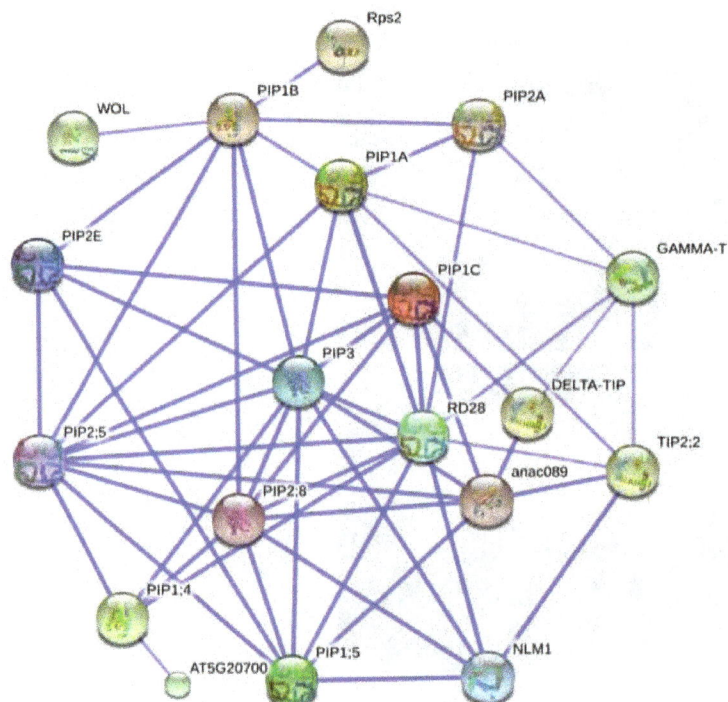

Fig. 4 Interactome of PIP1s subunits as obtained from string interaction. Interactome is shown in a confidence view where stronger associations are represented by thicker lines.

4 Discussion

Plant aquaporins is large family with at least 38 homologs, divided into four major subfamilies: plasma membrane intrinsic protein (PIP), tonoplast intrinsic protein (TIP), nodulin 26-like intrinsic proteins (NIP), and small and basic intrinsic proteins (SIP)[1]. The fourth subfamily of small and basic intrinsic proteins is not well characterized so far. Plasma membrane intrinsic protein (PIP) is divided into two groups: PIP1 and PIP2. In *Arabidopsis thaliana*, PIP1 has five homologs; PIP1a, PIP1b, PIP1c, PIP1; 4 and PIP1; 5 which are comprising six membrane-spanning domains tilted along the plane of the membrane and linked by five loops (A to E) located on the intra- (B, D) or extra-cytoplasmic (A, C, E) side of the membrane. The N- and C-terminal extremities are both exposed to the cytosol Fig. 3. A central aqueous pore is delineated by the transmembrane domains and loops B and E, which both carries a conserved Asn-Pro-Ala (NPA) motif and dip from either side of the membrane into the center of the molecule (Horton et al., 2007).

The functional interactome of PIP1A showed 7 interactions with PIP1B, PIP2; 5, PIP3, RD28 (PIP2C), TIP2; 2, GAMMA-TIP and PIP2A. The interactors: PIP2; 5, PIP3, RD28 have been confirmed experimentally by yeast-two hybrid system, assigning a new functions to PIP1A protein like osmo-regulation in plants under high osmotic stress, earlier examined under a high salt condition (Maurel et al., 2008). Other predicted interactor of PIP1A such as PIP1B is sharing water transport while TIP2; 2, GAMMA-TIP and PIP2A have the following functions; ammonia transporter/ methyl-ammonium transmembrane transporter/ water channel, facilitate the transport of water, diffusion of amino acids and/or peptides from the vacuolar compartment to the cytoplasm, play a role in the control of cell turgor and cell expansion and involved in root water uptake respectively (Zardoya and Villalba, 2001; Ishikawa et al., 2010; Braun et al., 2011).

PIP1B is essential for the water permeability of the plasma membrane and for the morphology of the root system. Our data showed that PIP1B is sharing 96% homology similarities of PIP1A and 3 interactors as well; PIP2A, PIP2; 5 and PIP3. These proteins had been experimentally confirmed by yeast-two hybrid as interactors of PIP1B (Maurel et al., 2008). The coordination between PIP1A and PIP1B in plant membranes regulates the osmotic pressure under salinity stress and control the cell turgor. Among the functional interactome of PIP1B, there are two interesting interactors; WOL (histidine kinase 4) is cytokinin-binding receptor that transducers cytokinin signals across the plasma membrane and Rps2 which is plasma membrane protein with leucine-rich repeat, leucine zipper, and P loop domains that confers resistance to *Pseudomonas syringae* infection by interacting with the virulence gene avrRpt2. RPS2 protein interacts directly with plasma membrane (Dortay et al., 2008; Kuwagata et al., 2012). Two other interactors of PIP1Bare PIP2E and PIP2; 8 having the common aquaporins function of water transport through plant cell membranes.

PIP1C and PIP1D are sharing 99.7% homology similarities, 3-D structure and functional interactome. The profile of their interactome comprises PIP2; 5, PIP3, RD28 and PIP2; 8 proteins which regulate the osmotic pressure under abiotic stress. Moreover, PIP1C interacts with anac089 protein (Arabidopsis NAC domain containing protein 89) which is negatively regulating floral initiation in Arabidopsis thaliana (Qi and Katagiri, 2009). PIP1C also interacts with DELTA-TIP; ammonia transporter/methylammonium transmembrane transporter which is the main channel of ammonia. It expresses especially in flowers, shoot and stem. PIP1; 3, PIP1; 5 and DELTA-TIP interacts with anac089 protein, involved in transcription factor activity (Li et al., 2010; Li et al., 2011). In addition PIP2E interacts with PIP1C assigning the function of response to nematode plus active water channel. PIP2E induce signals to plant immune system as response to nematode infection.

PIP1; 5 protein, as all PIP1 members, regulates the water channel activity and response to salt stress. The interactome profile of PIP1; 5 comprises PIP2; 5, PIP3, RD28, PIP2; 4, anac089 and PIP2; 8 which are the same interactors of PIP1C. These proteins, together with PIP1; 5 and PIP1; 3 are the main PIP1 family interactome controlling many physiological processes in plant cell like; osmoregulation in plants under high osmotic stress such as under a high salt, response to nematode, facilitate the transport of water across cell membrane and regulation of floral initiation in *Arabidopsis thaliana* (Kaldenhoff et al., 2007). Additionally, PIP1; 5 has a strong interaction with NLM1 (arsenite transmembrane transporter), assigning the PIP1; 5 member to share the coordination of arsenite transport and tolerance. NLM1 also acts as water channel regulator, probably required to promote glycerol permeability and water transport across cell membranes (Kamiya et al., 2009).

AQPs are a protein network in plant cell integrating in all physiological processes. Uncovering this network will enable us to infiltrate to core of cellular and molecular levels of the cell complexity. The functional interactome of PIP1protein sub-family is the first step to comprehend this complexity. The subcellular localization of PIP1 homologs is the key to unlock the complexity of their functional interactome, giving more details about AQPs network global function and physiological processes which they are interfering and where?

Abbreviations
AQP: Aquaporin
MIP: major intrinsic proteins
TIP: Tonoplast intrinsic proteins
PIP: Plasma membrane intrinsic proteins
SoPIP: Spinaciaoleracea plasma membrane intrinsic proteins
ClustalW2: multiple sequence alignment program, version 4
MSA: Multiple Sequence Alignment
Pymol: program to obtain 3D structure of the target proteins
WoLF PSORT: Subcellular localization prediction for each protein
PDB: Protein data bank

Acknowledgment
Authors thank the board of trustees, International University of Sarajevo for financial and moral support to realize this work.

References
Braun P, Carvunis AR, Charloteaux B, et al. 2011. Evidence for network evolution in an Arabidopsis interactome map. Science, 333(6042): 601-607

Dortay H, Gruhn N, Pfeifer A, et al. 2008. Toward an interaction map of the two-component signalling pathway of *Arabidopsis thaliana*. Journal of Proteome Research, 7(9): 3649-3660

Chaumont F, Moshelion M, Mark J. 2000. Daniels. Regulation of plant aquaporin activity. Franois Bilogy Cells, 97(10): 749-764

Hohmann I, Bill RM, Kayingo I, et al. 2000. Microbial MIP channels. Trends in Microbiology, 8: 33-38

Horton P, Park KJ, Obayashi T, et al. 2007. WoLF PSORT: Protein Localization Predictor. Nucleic Acids Research, 35: W585-W587

Franceschini A, Szklarczyk D, et al. 2013. STRING v9.1: protein-protein interaction networks, with increased coverage and integration. Nucleic Acids Research, 41(Database issue): D808-D815

Fraysse LC, Wells B, McCann MC, et al. 2005. Specific plasma membrane aquaporins of the PIP1 subfamily are expressed in sieve elements and guard cells. Biology of the Cell, 97(7): 519–534

Ishikawa H, Uenishi Y, Takase T, et al. 2010. Involvement of phytochrome A in regulation of water dynamics and aquaporin expression in Arabidopsis roots. 21st International Conference on Arabidopsis Research

Johanson U, Karlsson M, Johansson I, et al. 2001. The complete set of genes encoding major intrinsic proteins in Arabidopsis provides a framework for a new nomenclature for major intrinsic proteins in plants. Plant Physiology, 126: 1358-1369

Kaldenhoff R, Bertl A, Otto B, et al. 2007. Characterization of plant aquaporins. Methods in Enzymology, 428: 505-531

Kamiya T, Tanaka M, et al. 2009. NIP1;1, an aquaporin homolog, determines the arsenite sensitivity of Arabidopsis thaliana. Journal of Biological Chemistry, 284(4): 2114-2120

Kuwagata T, Ishikawa-Sakurai J, Hayashi H, et al. 2012. Influence of low air humidity and low root temperature on water uptake, growth and aquaporin expression in rice plants. Plant and Cell Physiology, 53(8): 1418-1431

Larkin MA, Blackshields G, Brown NP, et al. 2007. ClustalW and Clustal X version 2.0. Bioinformatics, 23(21): 2947-2948

Li J, Zhang J, Wang X, et al. 2010 A membrane-tethered transcription factor ANAC089 negatively regulates floral initiation in *Arabidopsis thaliana*. Science China Life Sciences, 1299-1306

Lamesch P, Berardini T.Z and Li D.2012.The Arabidopsis Information Resource (TAIR): improved gene annotation and new tools. Nucleic Acids Research, 40: 1202-1210

Letunic I, Doerks T, Bork P. 2012. SMART 7: recent updates to the protein domain annotation resource. Nucleic Acids Research, 40(1): 302-305

Maurel C, Verdoucq L, Luu DT, et al. 2008. Plant aquaporins: Membrane channels with multiple integrated functions. Annual Review of Plant Biology, 59: 595-624

Muller DJ, Janovjak H, Lehto T, et al. 2002. Observing structure, function and assembly of single proteins by AFM. Progress in Biophysics & Molecular Biology, 79: 1-43

Ping Li, Wind J, Shi XL, et al. 2011. Fructose sensitivity is suppressed in Arabidopsis by the transcription factor ANAC089 lacking the membrane-bound domain. Proceedings of the National Academy of Sciences of USA, 108: 3436–3441

Qi Y, Katagiri F. 2009. Purification of low-abundance Arabidopsis plasma-membrane protein complexes and identification of candidate components. Plant Journal, 57(5): 932-944

Robinson BH, Brooks RR, Kirkman JH, et al. 1996. Plant-available elements in soils and their influence on the vegetation over ultramar (serpentine) rocks in New Zealand. Journal of the Royal Society of New Zealand, 26: 457-468

Santoni V, Javot H, Lauvergeat V, et al. 2003. Role of a single aquaporin isoform in root water uptake. Plant Cell, 15: 509-522

Sayers E W, Barrett T, Benson S H. 2009. Database resources of the National Center for Biotechnology Information. Nucleic Acids Res, 37: 5-15.

Schwede T, Bordoli L, Kopp J, et al. 2006. The SWISS-MODEL Workspace: A web-based environment for protein structure homology modelling. Bioinformatics, 22: 195-201

The PyMOL Molecular Graphics System, Version 1.5.0.4. Schrödinger, LLC

Törnroth-Horsefield S, Wang Y, Hedfalk K, et al. 2005. Structural mechanism of plant aquaporin gating. Nature, 439(7077): 688-694

Wang Y, Tajkhorshid E. 2007. Molecular mechanisms of conduction and selectivity in aquaporin water channels. Journal of Nutrition, 137: 1509-1515

Yu J, Yool AJ, Schulten K, et al. 2006. Mechanism of gating and ion conductivity of a possible tetrameric pore in Aquaporin-1. Structure, 14: 1411-1423

Zardoya R. and Villalba S. 2001. A phylogenetic framework for the aquaporin family in eukaryotes. Journal of Molecular Evolution, 52(5): 391-404

Zelazny E, Borst JW, Muylaert M, et al. 2007. FRET imaging in living maize cells reveals that plasma membrane aquaporin interact to regulate their subcellular localization. Proceedings of the National Academy of Sciences of USA, 104: 12359-12364

8

Implementation of fuzzy system using different voltages of OTA for JNK pathway leading to cell survival/death

Shruti Jain[1], D.S. Chauhan[2]

[1]Department of Electronics and Communication Engineering, Jaypee University of Information Technology, Waknaghat, Solan, Himachal Pradesh 173234. India

[2]GLA University, Mathura, Uttar Pradesh, 281406. India

E-mail: jain.shruti15@gmail.com

Abstract

In this paper a well defined method for the design of JNK pathway for epidermal growth factor/ insulin using fuzzy system using operational transconductance amplifier was discussed. Fuzzy system includes fuzzification of the input variables, application of the fuzzy operator (AND or OR) in the antecedent, implication from the antecedent to the consequent, aggregation of the consequents across the rules, and defuzzfication. Fuzzy system with various electrical parameters for different voltages of OTA with different membership function was found. Results with 3V were the best.

Keywords epidermal growth factor (EGF)/ insulin; Jun N-terminal kinases (JNK); operational transconductance amplifier; fuzzy system; electrical parameters.

1 Introduction

Cell signaling pathways interact with one another to form networks. Such networks are complex in their organization and exhibit emergent properties such as bistability and ultra sensitivity. Analysis of signaling networks requires a combination of experimental and theoretical approaches including the development and analysis of models (Kevin et al., 2005; Gaudet et al., 2005; Weiss, 2001).

Bioelectronics encompasses a range of topics at the interface of biology and electronics. One aspect of the application of electronics in biology, medicine, and security includes both detection and characterization of biological materials, such as on the cellular and sub cellular level. Another aspect of bioelectronics is using biological systems in electronic applications (e.g., processing novel electronic components from DNA, nerves, or cells). Bioelectronics also focuses on physically interfacing electronic devices with biological systems (e.g., brain-machine, cell-electrode, or protein-electrode). Applications in this area include assistive technologies for individuals with brain-related disease or injury, such as paralysis, artificial retinas, and new technologies for

protein structure-function measurements. Fig. 1 shows the comparison of electronics and biological elements.

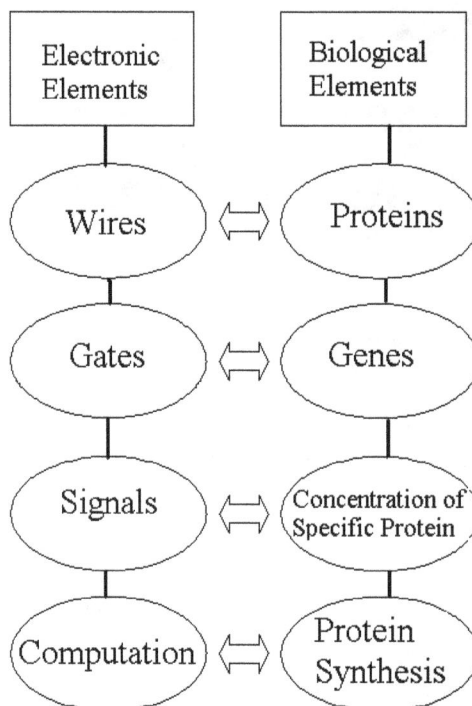

Fig. 1 Comparison of electronics and biological elements.

The decision between cell survival/ cell death is well regulated by three input signals : TNF, EGF and insulin. These factors in single or in combination activate various key players in the network pertaining to cell survival/ cell death. The epidermal growth factor (EGF) and EGF receptor (EGFR) were among the first growth factor ligand-receptor pairs discovered (Jain, 2012; Ullrich, 1990). Subsequently, EGFR was found to be a member of a receptor tyrosine kinase (RTK) family, the human epidermal growth factor receptor (HER) family (Arteaga, 2003; Jain et al., 2011). Each Insulin/Glucose Phospho-Antibody Array includes 85 highly specific and well-characterized phosphorylation antibodies in the Insulin/Glucose pathway. The epidermal growth factor receptor (EGFR) family plays an important role in cell lineage determination, the morphogenesis of many organs and in cell survival in the adult. Moreover, activating mutants and over-expression of these family members contribute to oncogenesis by inducing cells to proliferate and to resist cell death. Subsequent phosphorylation of the HER-kinase itself and/or other proteins, which then pass on to various signaling cascades (e.g., phosphoinositide 3-kinase (PI3K)/Akt mitogen-activated protein kinase (MAPK) pathways and, JAK/ STAT pathway), can lead to different cellular events such as growth, migration, and division. Insulin receptor substrate 1 (IRS-1) plays important biological function for both metabolic and mitogenic (growth promoting) pathways: mice deficent of IRS1 have only a mild diabetic phenotype, but a pronounced growth impairment, i.e. IRS-1 plays a key role in transmitting signals from the insulin and insulin-like growth factor-1 (IGF-1) receptors to intracellular pathways PI3K / Akt and Erk MAP kinase pathways.

MAP kinases are actually a family of protein kinases that are widely distributed and are found in all eukaryotic organisms. These can be classified into three main functional groups (Jain et al., 2010; Jain, 2014a, b, c). The first is mediated by mitogenic and differentiation signals i.e., extracellular signal regulated kinase (ERK) pathway, other respond to stress and inflammatory cytokines i.e., *Jun N-terminal kinases* (JNK)/ stress activation protein kinase (SAPK) pathway and the third pathway is p38/HOG pathway HOG stands for high osmolarity glycerol where the p38 proteins are a subfamily. Each of these pathways led to the dual phosphorylation of MAP kinase family members responsible for activation of transcription factors.

This paper presents the implementation of JNK pathway using fuzzy system using operational transconductance amplifier (OTA) with different voltages. There are five parts of the fuzzy system (Jain, 2014a, b, c). Fuzzification of the input variables, application of the fuzzy operator (AND or OR) in the antecedent, implication from the antecedent to the consequent, aggregation of the consequents across the rules, and defuzzfication.

Operational Transconductance Amplifier (OTA) is a voltage controlled current source (VCCS) (Jain, 2014a, b, c). The input stage will be a CMOS differential amplifier. Since the output resistance of the differential amplifier is reasonably high. If both higher output resistance and more gain are required, then the second stage could be a cascade with a cascade load. The output is current; to convert current to voltage we can use current mirror circuit is used at the output side of VCCS circuit. Current mirror circuit increases the gain of the circuit.

This paper presents the electrical parameters i.e. differential input resistance, output resistance, large signal voltage gain (20 $\log_{10} V_o/V_{id}$ in dB), common mode rejection ratio (CMRR) (20 $\log(A_d / A_{cm})$ in dB), slew rate (SR) (max (dV_o / dt) in V/µsec). For simulation and calculation of parameters I have used SPICE software (Rashid, 2009).

2 Signaling Pathway of Egf/ Insulin

Epidermal growth factor (EGF) / Insulin is a growth factor that plays an important role in the regulation of cell growth, proliferation, and differentiation. It also increases cancer risk. EGF acts by binding with high affinity to epidermal growth factor receptor (EGFR) on the cell surface and stimulating the intrinsic protein-tyrosine kinase activity of the receptor (see the second diagram). The tyrosine kinase activity, in turn, initiates a signal transduction cascade that results in a variety of biochemical changes within the cell - a rise in intracellular calcium levels, increased glycolysis and protein synthesis, and increases in the expression of certain genes including the gene for EGFR / IRS-1 that ultimately lead to DNA synthesis and cell proliferation (Jain, 2012; Jain, 2014a, b, c).

Upon ligand-binding receptors homo-dimerise or hetero-dimerise triggering tyrosine trans-phosphorylation of the receptor sub-units. These tyrosine phosphorylated sites allow proteins to bind through their Src homology 2 (SH2) domains leading to the activation of downstream signaling cascades including the RAS/extracellular signal regulated kinase (ERK) pathway, the phosphatidylinositol 3 kinase (PI3K) pathway and the activator of transcription (JAK/ STAT) pathway. Differences in the C-terminal domains of the ErbB receptors govern the exact second messenger cascades that are elicited conferring signaling specificity. The EGF signal is terminated primarily through endocytosis of the receptor-ligand complex. The contents of the endosomes are then either degraded or recycled to the cell surface. A number of signal transduction pathways branch out from the receptor signaling complex as shown in Fig. 2.

The most widely studied MAP kinase cascade is the JNK/SAPK (c-Jun NH_2-terminal kinase/stress activated protein kinase). The c-Jun kinase (JNK) is activated when cells are exposed to ultraviolet (UV) radiation, heat shock, or inflammatory cytokines. However, the functional consequence of JNK activation in UV-irradiated cells has not been established. The absence of JNK caused a defect in the mitochondrial death signaling pathway, including the failure to release cytochrome c (Jain, 2012).

3 Methodology

Our aim is to design and implementation of fuzzy system using Operational transconductance amplifier (OTA) for JNK pathway of EGF/Insulin. The problem is to estimate cell survival or death for JNK (one of the path of MAPK) pathway used in EGF. Let's assume inputs as SOS and JNK and output as state of cell. Define the linguistic variables to all input and output. The linguistic variables defined for *SOS* are RAF, RAL and p38, for *JNK* (*MAPK*) is absent, not lying and present while, for state of cell are cell survival, no function and cell death. Assign membership functions (S, Z, triangular and trapezoidal) to every linguistic variable. I have designed all membership functions using OTA (Jain, 2014a, b, c).

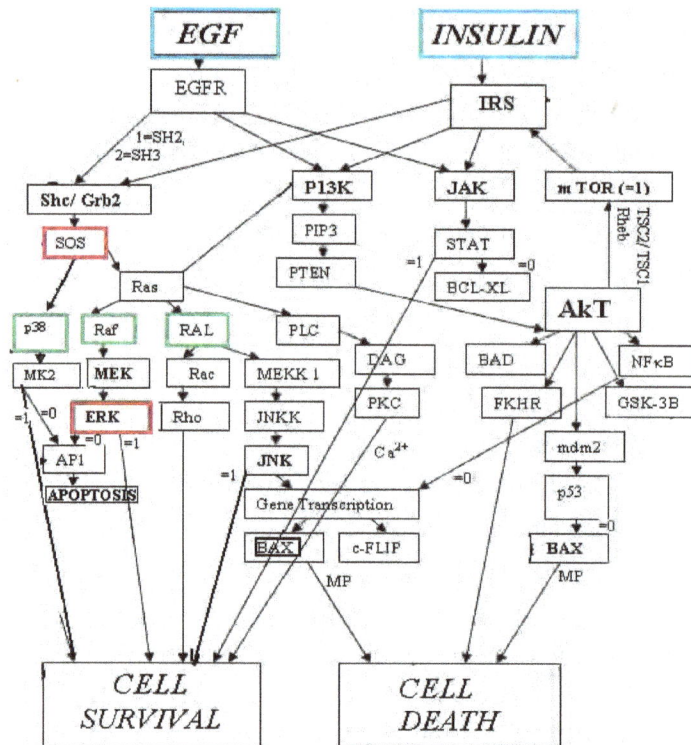

Fig. 2 Illustration of signal communication network triggered by EGF/ Insulin.

Assign different membership functions to all linguistic variables. Then, we define the range of each linguistic variable: absent as 0-4, not lying as 3-7 and present as 6-10.Similarly, for every input and output function. Let's define these ranges in volts so as to implement electronically: absent as $1V$, not lying as $2V$ and present as $3V$. Similarly, RAF as $1V$, RAL as $2V$ and p38 as $3V$, and cell survival as $1V$, cell death as $2V$ and no function as $3V$. I have assumed reference voltage as $4V$ and the OTA works at $4V$.

Second step was *rule composition* i.e. IF-THEN statement. There are different operators like AND (min), OR (max) and inverter. I have designed all the operators using OTA (Jain, 2014a, b, c).

The rules are:

- IF SOS is RAF *AND* JNK is not lying THEN state of cell is no function.
- IF SOS is RAL *AND* JNK is present THEN state of cell is survival.
- IF SOS is RAL *AND* JNK is absent THEN state of cell is death.

Before THEN and after IF is known as antecedent part and after THEN is consequent part. Antecedent part is rule composition part and consequent part is *implication process*. Third step implication process is of different types but in this paper I have used mamdani implication style. Electronically rules are used as

- IF SOS is RAF (1V) *AND* JNK is not lying (2V) THEN state of cell is no function (3V).
- IF SOS is RAL (2V) *AND* JNK is present (3V) THEN state of cell is survival (1V).
- IF SOS is RAL (2V) *AND* JNK is absent (1V) THEN state of cell is death (2V).

Similarly we can make different rules. I have implemented only three rules electronically and, get their output. Fourth step is to *aggregate* (add) the every output after implication process and then final step is to *defuzzify* it. In this paper I have used Maximum defuzzification techniques.

Fig. 3 shows the output after every step. V(32), V(62) and V(89) are the output after mamdani implication process, V(36) is output after aggregating all the rules which we are getting after implication process. V(65) is the final output i.e defuzzified output.

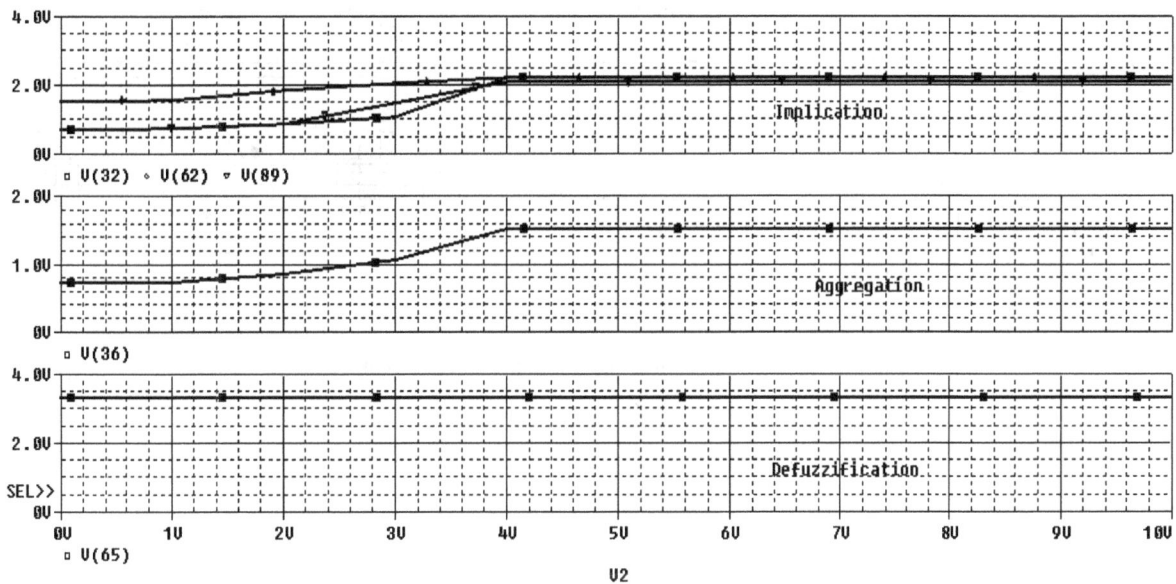

Fig. 3 Final output of Fuzzy System using 6V.

Let's vary the OTA voltages from 3V to 6V. Aspect ratio (W/L) is calculated for every voltage value shown in Table 1.

Table 1 Different values of W/L for different OTA voltages.

OTA voltage ⟶	6V	5V	4V	3V
$(W/L)_1 = (W/L)_2$	41.42	18.4	27.6	20.7
$(W/L)_3 = (W/L)_4$	0.125	0.22	0.75	4
$(W/L)_5 = (W/L)_6$	0.247	0.16	0.232	0.291
$(W/L)_7 = (W/L)_8$	0.518	0.336	0.48	0.611
I_D (μA)	50	50	75	100

Fig. 4, Fig. 5 and Fig. 6, shows the fuzzy system using different voltages of OTA i.e. 5V, 4V and 3V respectively. The output after every step i.e. V(32), V(62) and V(89) are the output after mamdani implication process, V(36) is output after aggregating all the rules which we are getting after implication process. V(65) is the final output, i.e., defuzzified output.

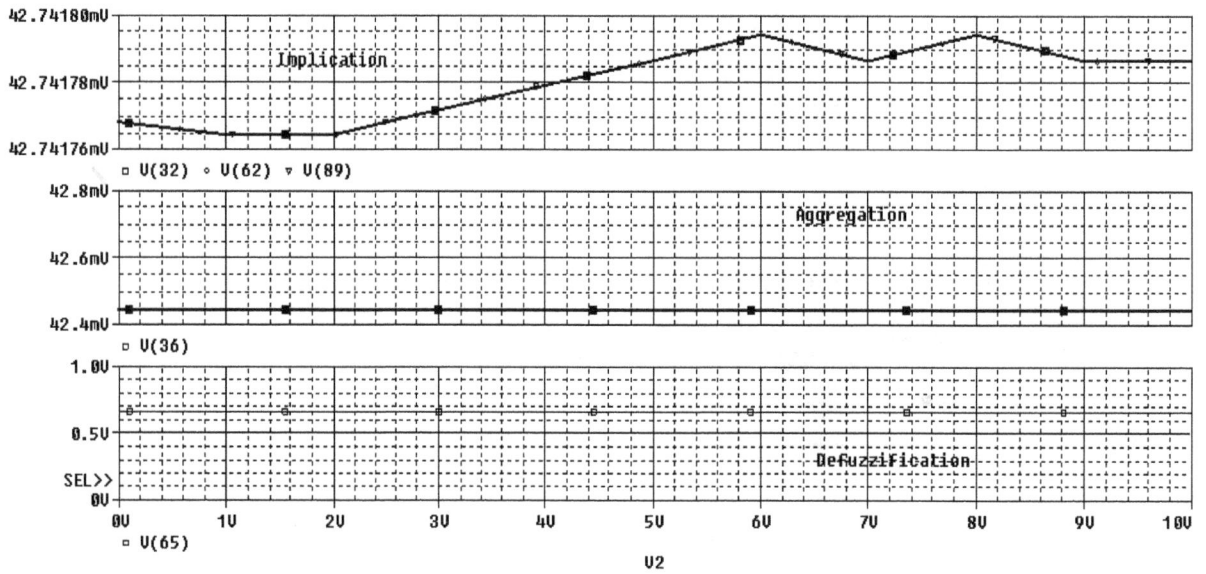

Fig. 4 Final output of fuzzy system using 5V.

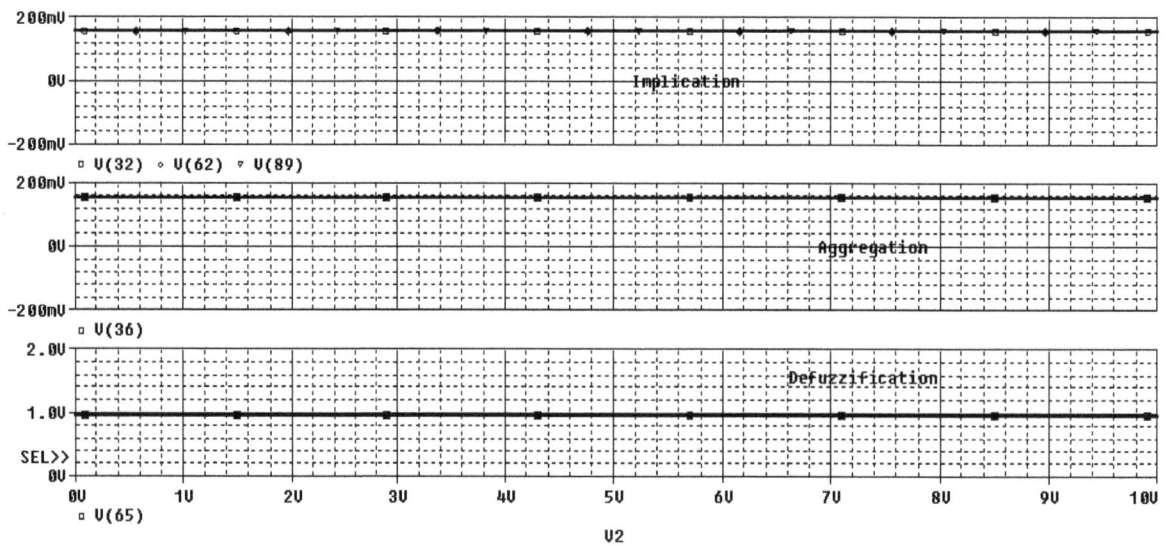

Fig. 5 Final output of fuzzy system using 4V.

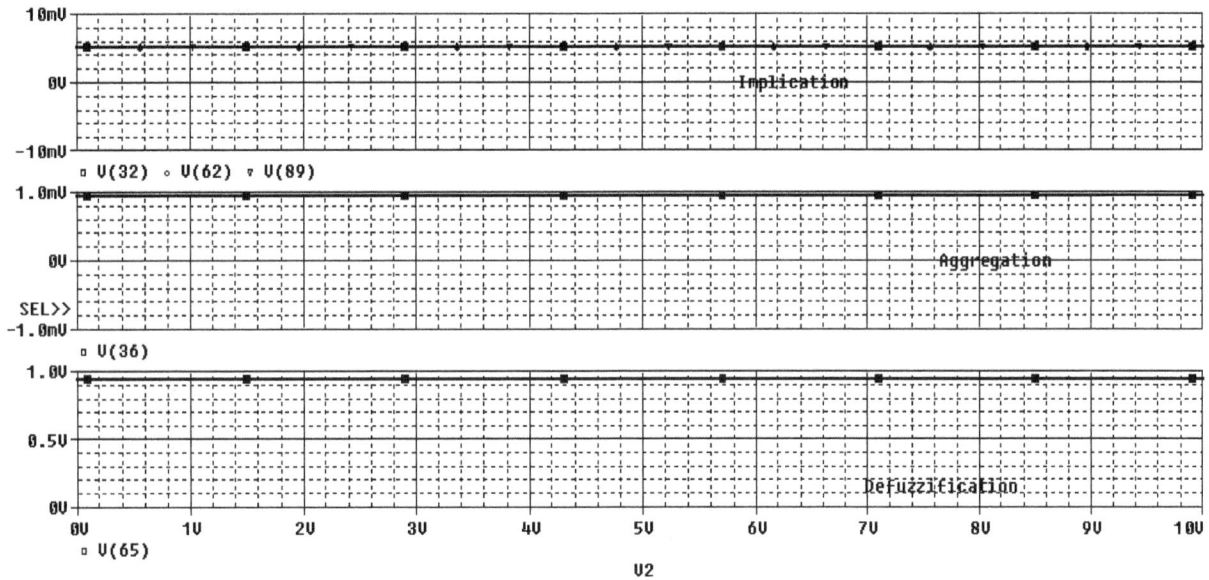

Fig. 6 Final output of fuzzy system using 3V.

Consider the different membership functions as input. Table 2 (includes only S and Z membership function) and Table 3 (S, Z, triangular and trapezoidal membership function) show the comparison of electrical parameters of balanced OTA. Results using 3V and all membership functions are the best. In this I have considered all (S, Z, triangular and trapezoidal membership function) membership functions for various linguistic variables instead of only one or two.

Table 2 Comparison of the electrical parameters of fuzzy system using signoidal (S) and anti-sigmoidal (Z) membership function only.

	Balanced OTA using S and Z Membership functions (6V)	Balanced OTA using S and Z Membership functions (5V)	Balanced OTA using S and Z Membership functions (4V)	Balanced OTA using S and Z Membership functions (3V)
Voltage gain (dB)	42.325	51.77	53.095	61.49
Input resistance (KΩ)	45	46	47.5	55
Output resistance (KΩ)	66.6	196	199	195
CMRR (dB)	18.02	20.47	23.65	27.60
Slew Rate (V/μ sec)	0.5	0.4	0.45	0.6
Power dissipation (mW)	0.217	2.18	3	3.26

Table 3 Comparison of the electrical parameters of fuzzy system using any S, Z, triangular and trapezoidal membership function.

	OTA using S, Z, triangular and trapezoidal Membership functions (6V)	OTA using S, Z, triangular and trapezoidal Membership functions (5V)	OTA using S, Z, triangular and trapezoidal Membership functions (4V)	OTA using S, Z, triangular and trapezoidal Membership functions (3V)
Voltage gain (dB)	219.38	514.77	534.095	615.49
Input resistance (KΩ)	61.4	46	47.5	55
Output resistance (KΩ)	66.67	196	199	195
CMRR (dB)	95.98	20.47	23.65	27.60
Slew Rate (V/μ sec)	0.5	0.4	0.45	0.6
Power dissipation (mW)	0.282	2.18	3	3.26

The OTA is a current source, and the output impedance of the device is high, in contrast to the op-amp's very low output impedance. To increase the speed of the system, CMRR should be high. In our calculations CMRR is very high.

4 Conclusion

The paper gives the designing of the JNK pathway of EGF/ insulin using fuzzy system using OTA. I have successfully designed and implemented all the steps of fuzzy system i.e. fuzzification, rule composition, implication, aggregation and defuzzification process for various rules. I have also calculated its various parameters like slew rate, CMRR, power dissipation, gain, input resistance and output resistance using S and Z membership function as one and all membership functions as second with different voltages of OTA. The results with 3V of OTA and different membership functions are the best.

Abbreviations

ASK1, Apoptosis signal-regulating kinase 1; CMRR, Common mode rejection ratio; EGF, epidermal growth factor; EGFR, epidermal growth factor receptor; ERK, extracellular-regulated kinase; Grb2, growth factor receptor-bound 2;   IGF, insulin-like growth factor; IκB, I Kappa B (nuclear factor of kappa light polypeptide gene enhancer in B-cells inhibitor); IKK, IκB kinase; IR, insulin receptor; IRS1, insulin receptor substrate 1; JNK1, c-jun NH$_2$ terminal kinase 1; MAP kinases, mitogen-activated protein kinases; MEK, mitogen-activated protein kinase and extracellular-regulated kinase kinase; MK2, mitogen-activated protein kinase-activated protein kinase 2; NF-κB, nuclear factor-κB; OTA, Operational Transconductance amplifier; PLADD, pre-ligand assembly domain; PI3K, phosphatidylinositol 3-kinase; p38, P38 mitogen-activated protein kinases; SAPK/JNK , Stress-activated protein kinase/Jun-amino-terminal kinase; SH2, Src homolgy 2; SOS, Son of Sevenless.

References

Allen PE, holberg DR. 2011. CMOS Analog Circuit Design. International Student Edition, Oxford, UK

Arteaga C. 2003. Targeting HER1/EGFR: a molecular approach to cancer therapy Seminars in Oncology, 30: 314

Gaudet S, Kevin JA, John AG, et al. 2005. A compendium of signals and responses trigerred by prodeath and prosurvival cytokines. Manuscript M500158-MCP200.

Jain S, Bhooshan SV, Naik PK. 2010. Model of mitogen activated protein kinases for cell survival/death and its equivalent bio-circuit. Current Research Journal of Biological Sciences, 2(1): 59-71

Jain S, Bhooshan SV, Naik PK. 2011. Mathematical modeling deciphering balance between cell survival and cell death using insulin. Network Biology, 1(1): 46-58

Jain S. 2012. Communication of signals and responses leading to cell survival / cell death using Engineered Regulatory Networks. PhD Thesis. Jaypee University of Information Technology, Solan, Himachal Pradesh, India

Jain S. 2014a. Design and simulation of fuzzy membership functions for the fuzzification module of fuzzy system using operational amplifier. International Journal of Systems, Control and Communications, 6(1): 69-83

Jain S. 2014b. Design and simulation of fuzzy implication function of fuzzy system using two stage CMOS operational amplifier. International Journal of Emerging Technologies in Computational and Applied Sciences, 7(2): 150-155

Jain S. 2014c. Implementation of fuzzy system using operational transconductance amplifier for ERK pathway of EGF/ Insulin leading to cell survival/ death. Journal of Pharmaceutical and Biomedical Sciences, 4(8): 701-707

Kevin JA, John AG, Gaudet S, et al. 2005. A systems model of signaling identifies a molecular basis set for cytokine-induced apoptosis. Science, 310: 1646-1653

Rashid MH. 2009. Introduction to PSICE Using OrCAD for Circuits and Electronics (3rd edition). PHI, India

Ullrich A, Schlessinger J. 1990. Signal transduction by receptors with tyrosine kinase activity. Cell, 61: 203-211

Weiss R. 2001. Cellular computation and communications using engineered genetic regulatory networks. PhD Thesis. MIT, USA

9

A bioinformatics and network analysis framework to find novel therapeutics for autoimmunity

Soumya Banerjee[1,2]
[1]University of Oxford, Oxford, United Kingdom; [2]Ronin Institute, Montclair, USA
E-mail: soumya.banerjee@maths.ox.ac.uk

Abstract

The immune system protects a host from foreign pathogens. In rare cases, the immune system can attack the cells of the host organism causing autoimmune diseases. We outline a computational framework that combines bioinformatics and network analysis with an emerging targets platform. The computational framework presented here can be used to find drug targets for autoimmune diseases. It can also be used to find existing drugs that can be repurposed to treat autoimmune diseases based on networks of interactions or similarities between different diseases. Information on which gene regions are associated with the disease (single nucleotide polymorphisms) can be used in gene therapy when that technique becomes viable. Our analysis also revealed immune cell subtypes that are implicated in these diseases. These immune cell subtypes can be selected for immunotherapy experiments. Finally, our analysis also reveals intra-cellular and protein-protein interaction networks and pathways that can be targeted with small molecule inhibitors. The downstream off-target effects of these inhibitors can also be determined from such a network analysis. In summary, our computational framework can be used to find novel therapeutics for autoimmune diseases and potentially even other dysfunctions.

Keywords autoimmune diseases; bioinformatics; network analysis; immune system modelling; agent based models; ordinary differential equation models.

1 Introduction

The immune system is tasked with protecting a host from external pathogens (Takutu et al., 2011; Jesmin et al., 2016). It is trained to not recognize self (peptides that are expressed by normal cells in the body of the host). However, in some cases it erroneously recognizes and attacks normal cells in the host. These are called autoimmune diseases.

Insights from bioinformatics coupled with data from emerging curated repositories can lead to novel therapeutics that mayameliorate symptoms of these diseases and help patients lead a normal lifestyle (Zhang, 2016a, b).

We use bioinformatics and network analysis techniques combined with an emerging drug targets platform. We use this approach to show how insights can be derived into potential drug targets of two autoimmune diseases: systemic lupus erythematosus and Sjogren disease. Our work shows the potential of combining computational techniques with emerging repositories to drive insights into the immune system in health and disease.

2 Methods

Our approach is to combine bioinformatics and network analysis approach with an emerging platform that has curated information on diseases and their drug targets. We use an emerging platform to quantify targets for an autoimmune disease (https://www.targetvalidation.org/) (Koscielny et al., 2017).

Our computational framework combines bioinformatics with network approaches along with a novel repository. Such an approach can enable search for new targets for diseases and insights into mechanisms. We show this approach here for autoimmune diseases (Fig. 1). Our code is available for download from a repository (https://bitbucket.org/neelsoumya/autoimmune_targets_pipeline).

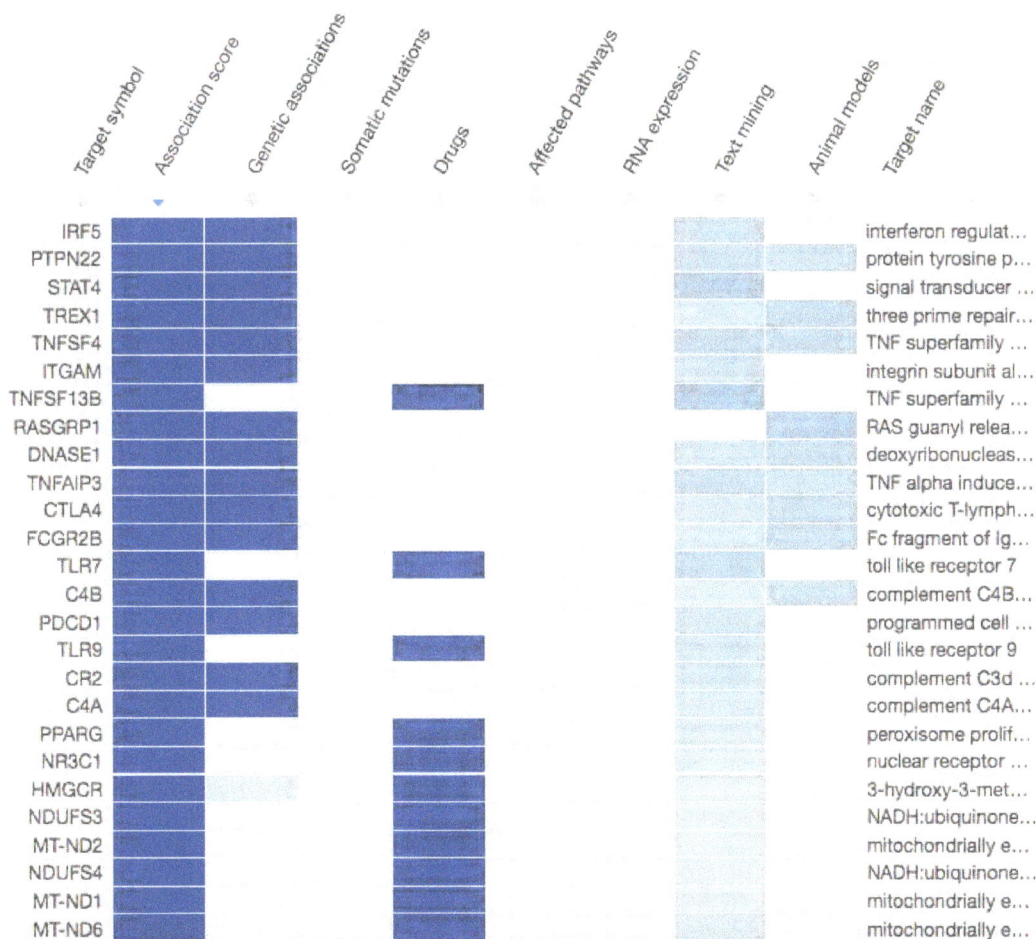

Fig. 1 A depiction of the top targets for systemic lupus erythematosus. IRF5 is predicted to be a very important target along with other factors (available from https://www.targetvalidation.org/disease/EFO_0002690/associations and https://www.targetvalidation.org/disease/EFO_0002690).

3 Results

3.1 Top targets associated with systemic lupus erythematosus

Querying the platform revealed that Interferon Regulatory Factor 5 (IRF5) is a top factor involved in systemic lupus erythematosus.

Some of the known drugs in use or currently in testing for systemic lupus erythematosus are shown in Table 1. Most of the drugs are small molecule inhibitors. A complete list is available in Supplementary Information.

Table 1 Known drugs in use or testing for systemic lupus erythematosus (top 10 drugs; complete list available in Supplementary Information).

Drug	Type	Mechanism of action
DEXAMETHASONE	N/A	Small molecule
DEXAMETHASONE	N/A	Small molecule
METFORMIN	Recruiting	Small molecule
METFORMIN	Recruiting	Small molecule
PIOGLITAZONE	Completed	Small molecule
DEXAMETHASONE PHOSPHORIC ACID	N/A	Small molecule
PREDNISOLONE	N/A	Small molecule
PREDNISONE	N/A	Small molecule
METFORMIN	Recruiting	Small molecule

This kind of information can be used to find novel drug targets.

3.2 IRF5 interaction network

We also constructed the network of interactions between interferon regulatory factor 5 (IRF5) and other factors (Fig. 2).

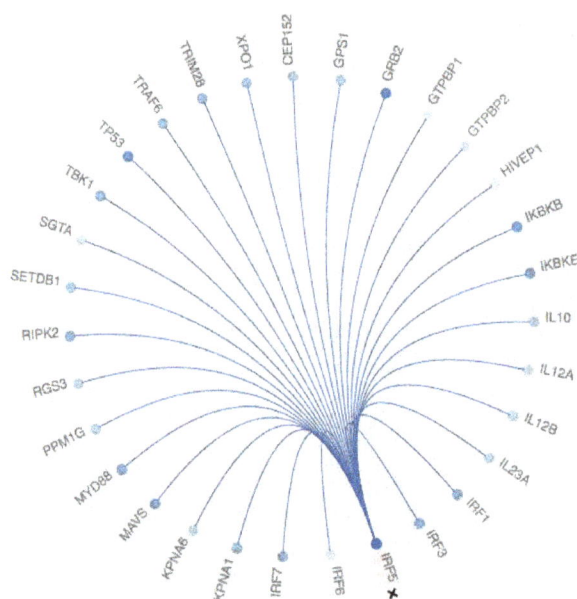

Fig. 2. Network of interactions between Interferon regulatory factor 5 and other factors (available from https://www.targetvalidation.org/target/ENSG00000128604).

This kind of network of interactions with other factors can be used to find drug targets that can influence IRF5. IRF5 is a very important factor that is also implicated in a host of other autoimmune diseases. We show the association of IRF5 with other diseases like rheumatoid arthritis and inflammatory bowel disease (Fig. 3).

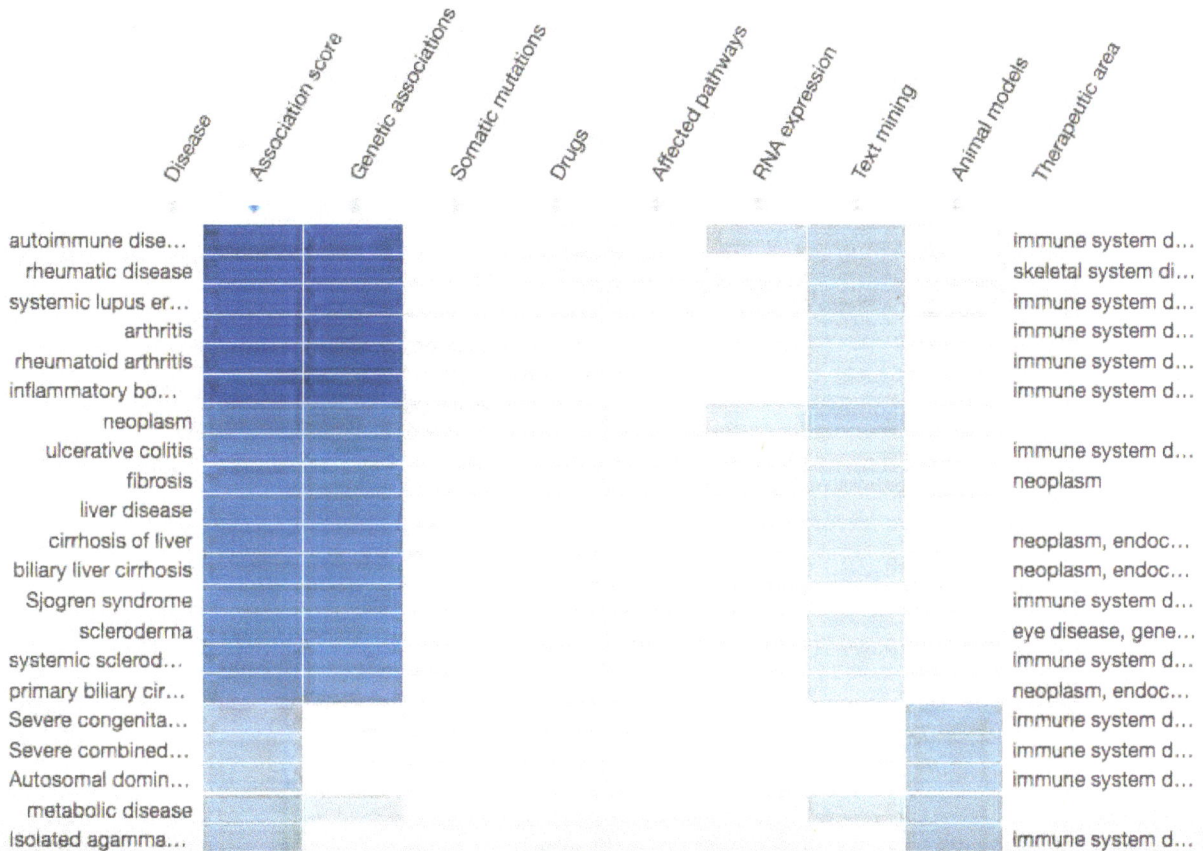

Fig. 3 IRF5 is a very important factor that is also implicated in a host of other autoimmune diseases. The diseases are shown in this figure (available from https://www.targetvalidation.org/target/ENSG00000128604/associations?view=t:table).

3.3 Interactions with other diseases: Sjogren syndrome

We demonstrate our approach by using another autoimmune disease called Sjogren syndrome. We look at the network of associations between other diseases and Sjogren syndrome (Fig. 4). This kind of information can be used to find existing drugs that can be repurposed to treat this disease.

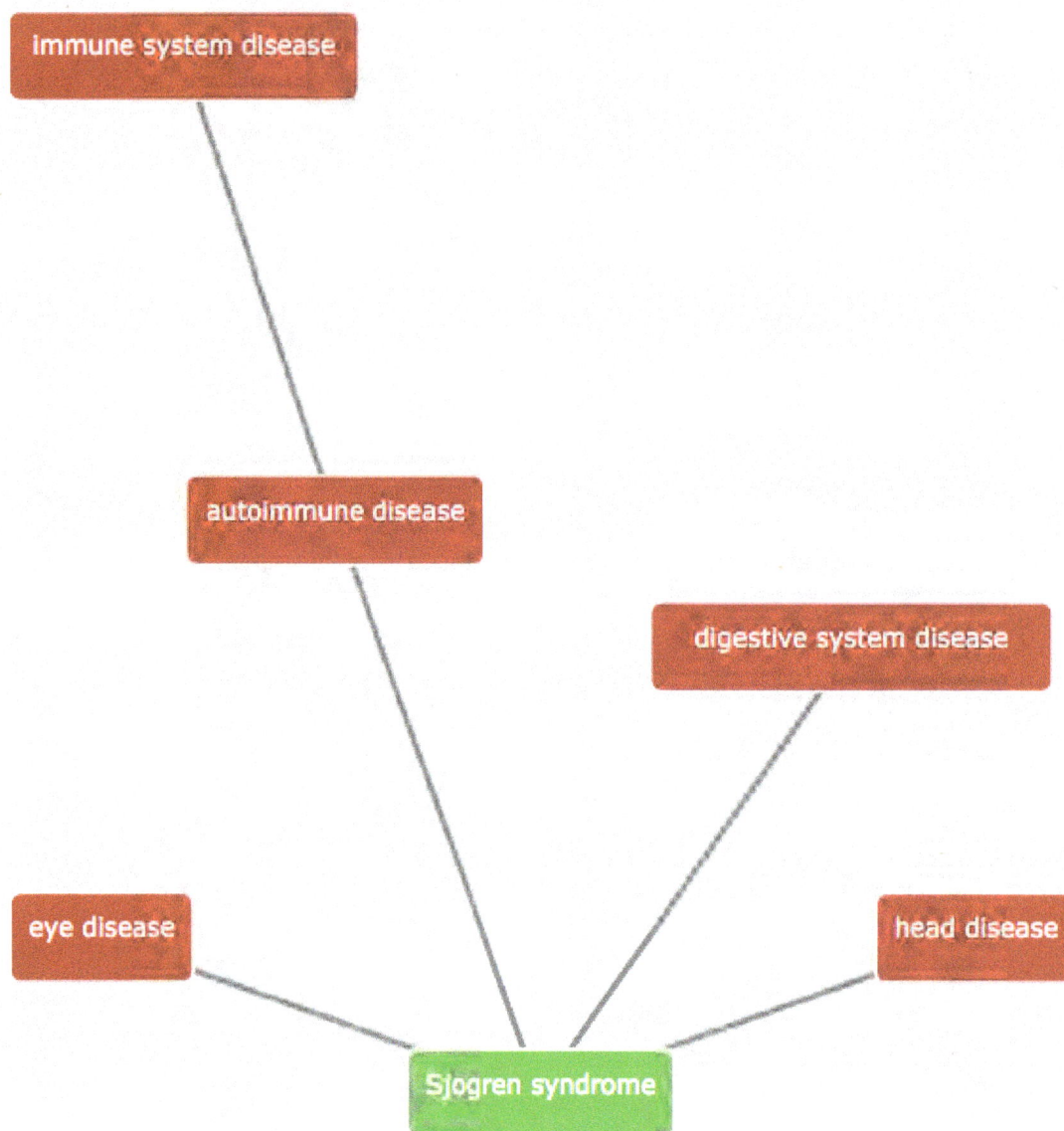

Fig. 4 A network of associations between other diseases and Sjogren syndrome (available from https://www.targetvalidation. org/disease/EFO_0000699).

3.4 Targets and associations

The associations of other diseases with Sjogren disorder are shown in Fig. 5.

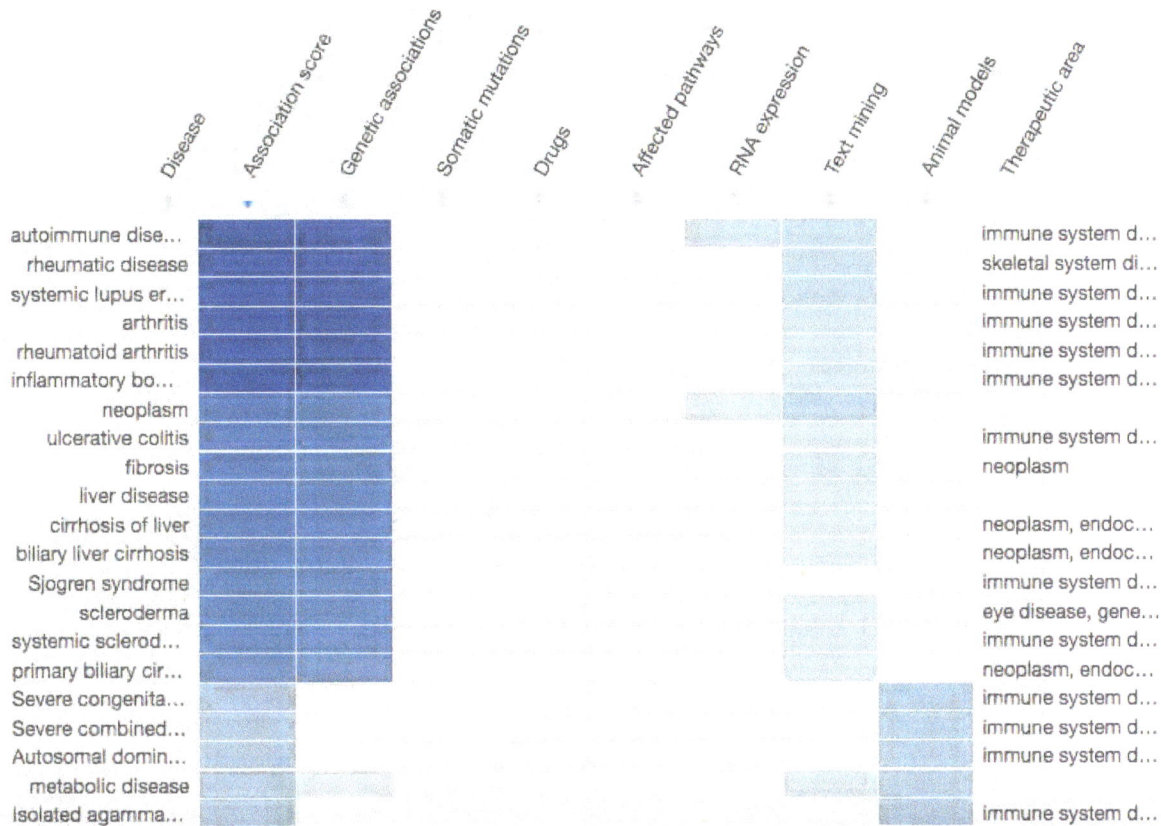

Fig. 5 Associations of Sjogren disorder with other diseases (available from https://www.targetvalidation.org/disease/EFO_0000699/associations).

Some of the current drugs that are in use or development for Sjogren syndrome are also shown in Table 2.

Table 2 Drugs that are in development or use for Sjogren disorder.

Drug	Status	Type
DEXAMETHASONE	Completed	Small molecule
HYDROXYCHLOROQUINE	Completed	Small molecule
HYDROXYCHLOROQUINE	Completed	Small molecule
BAMINERCEPT	Terminated	Protein
MYCOPHENOLATE MOFETIL	Enrolling by invitation	Small molecule
EFALIZUMAB	Terminated	Antibody
BAMINERCEPT	Terminated	Protein
BELIMUMAB	Completed	Antibody
MYCOPHENOLATE MOFETIL	Enrolling by invitation	Small molecule
BELIMUMAB	Recruiting	Antibody
EFALIZUMAB	Terminated	Antibody
RITUXIMAB	Recruiting	Antibody
RITUXIMAB	Completed	Antibody
TOCILIZUMAB	Recruiting	Antibody
RITUXIMAB	Completed	Antibody

3.5 IRF5 intra-cellular pathway and interaction network

We constructed the network of IRF5 intra-cellular pathway and interaction networks (Fig. 6). These network diagrams can be used to find more novel targets.

The network diagrams suggest there are key hubs. This can be used to find off-target effects or side effects of drugs by outlining key connections to other functions.

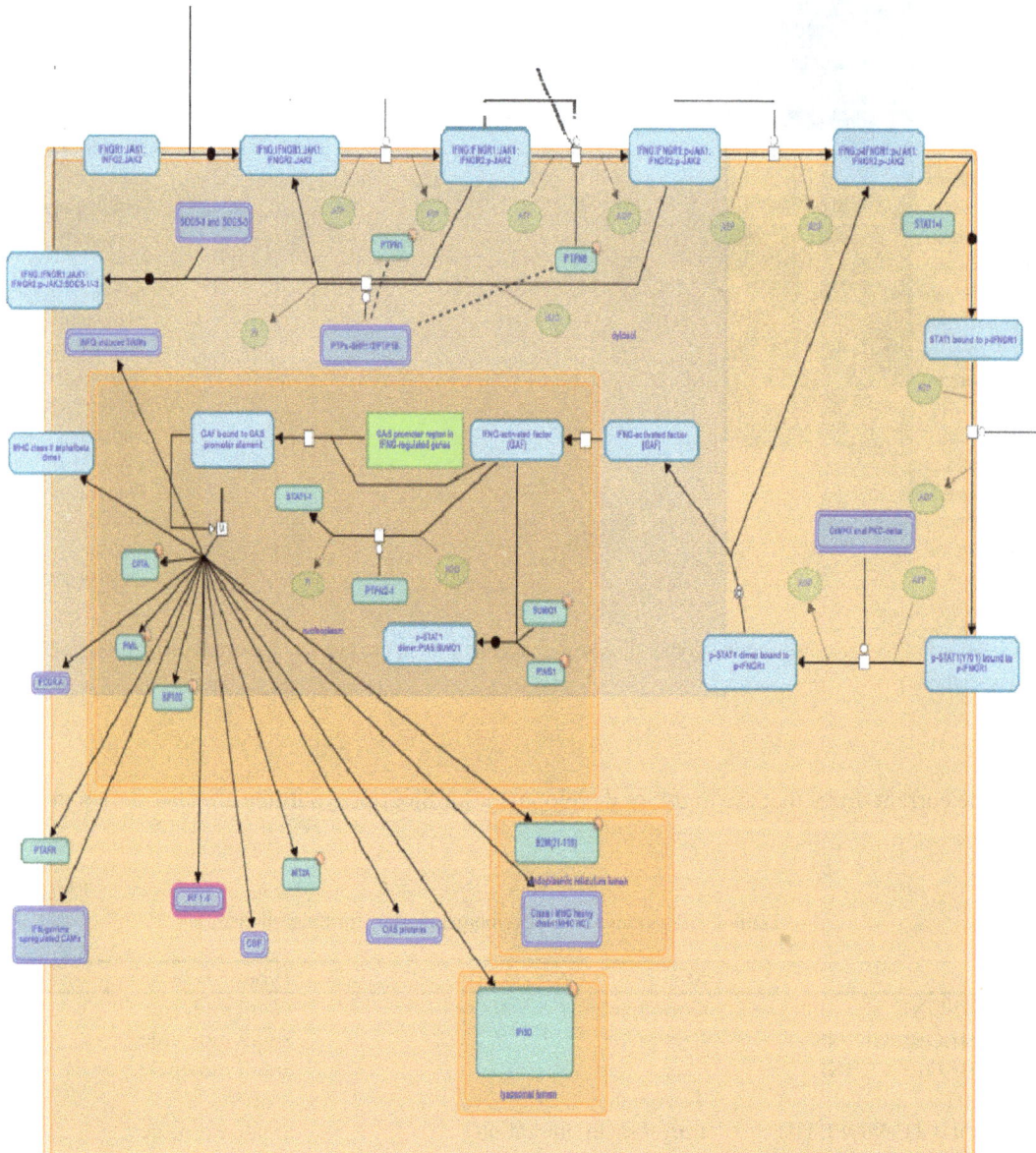

Fig. 6 A network of IRF5 intra-cellular pathway and interaction networks.

3.6 Genes and single nucleotide polymorphisms associated with disease

Using the computational framework, we found single nucleotide polymorphisms (SNP) in genes associated with these diseases. One of these is rs2004640 which is a SNPin the IRF5 gene in the chromosomal region 7q32.1. This is associated with systemic lupus erythematosus. Some of the genes and transcript consequences are listed in Table 3. A full list is available online (see supplementary Information; filename: Mappings-Homo _sapiens_Variation_Mappings_rs2004640_consequences.csv).

Table 3 Tenes and transcript consequences for systemic lupus erythematosus (full list available at http://www.ensembl.org/ Homo_sapiens/ Variation/Mappings?db=core;r=7:128937747-128938747;v=rs2004640;vdb=variation;vf=1430080)

Gene	Consequence Type
ENSG00000128604 HGNC: IRF5	splice donor variant
ENSG00000128604 HGNC: IRF5	intron variant
ENSG00000128604 HGNC: IRF5	intron variant
ENSG00000128604 HGNC: IRF5	intron variant non coding transcript variant
ENSG00000128604 HGNC: IRF5	intron variant
ENSG00000128604 HGNC: IRF5	upstream gene variant
ENSG00000128604 HGNC: IRF5	upstream gene variant
ENSG00000128604 HGNC: IRF5	upstream gene variant
ENSG00000128604 HGNC: IRF5	intron variant NMD transcript variant
ENSG00000128604 HGNC: IRF5	intron variant

Some of the transcripts with tissues where they are likely expressed in are listed in Table 4. A full list is available online (see Supplementary Information; filename: Mappings-Homo_sapiens_Variation_Mappings_ rs2004640.csv).

Table 4 Transcripts with tissues where they are likely expressed and effect size (for SNP in rs2004640).

Gene	Effect size	Tissue
ENSG00000128594	0.002135703	Adipose_Subcutaneous
ENSG00000135245	0.017383528	Adipose_Subcutaneous
ENSG00000240758	0.016198117	Adipose_Subcutaneous
ENSG00000128524	0.019061794	Skin_Sun_Exposed_Lower_leg
ENSG00000128594	-0.118246307	Thyroid
ENSG00000230715	-0.023401133	Adipose_Subcutaneous
ENSG00000243679	0.002803941	Adipose_Subcutaneous
ENSG00000229413	-0.061623932	Adipose_Subcutaneous
ENSG00000271553	-0.070702285	Adipose_Subcutaneous
ENSG00000205085	-0.124847132	Adipose_Subcutaneous

We note that some of the genes (like ENSG00000128594) are associated with thyroid disease, which is observed in patients with systemic lupus erythematosus. Finally, one of the regulatory features (ENSR00000217801) is active in cell lines consisting of many immune system cells (like B cells, natural killer cells and macrophages), and normal cells like those found in the spleen and pancreas (full list available from Supplementary Information; filename: Mappings-Homo_sapiens_Variation_Mappings_rs2004640_ regulatory.csv). In the future, it may be possible to use therapeutics to specifically target these cells or sites.

4 Discussion

The immune system protects the host against foreign pathogens. In rare circumstances, it can harm cells of the host causing autoimmune diseases. We present a computational framework that combines bioinformatics and network analysis approaches with data from a novel platform. Our framework can be used to find novel therapeutic strategies for autoimmune diseases. We show it using two autoimmune diseases: systemic lupus erythematosus and Sjogren disorder.

The computational framework presented here can be used to find drug targets for autoimmune diseases. It can also be used to find existing drugs that can be repurposed to treat autoimmune diseases. For example, Sjogren is associated with other diseases (Fig. 5). Association information of this nature can be used to repurpose existing drugs to treat these diseases.

Information on which gene regions are associated with the disease (single nucleotide polymorphisms in Table 3) can be used in gene therapy when these techniques become viable.

Our analysis also reveals intra-cellular and protein-protein interaction networks and pathways that can be targeted with small molecule inhibitors. The downstream off-target effects of these inhibitors can also be determined from such a network analysis (intra-cellular regulatory network for IRF5, Fig. 6).

Our framework also revealed immune cell subtypes and specific sites that are implicated in these diseases (associated with a regulatory feature: ENSR00000217801). These immune cell subtypes can be selected for immunotherapy experiments. In the future, it may also be possible to use therapeutics to specifically target these sites instead of using systemic drugs.

Our approach can be combined with infectious disease models (Banerjee, 2013; Banerjee, 2015a, b; Banerjee and Moses, 2010; Banerjee et al., 2016). Population level approaches like ordinary differential equations are computationally tractable and can scale up to simulate host pathogen dynamics in large organisms (Banerjee and Moses, 2009). These can also be used to investigate the role of molecular mimicry in autoimmune diseases. Finally, hybrid modelling approaches can also be very useful in modelling these biological systems (Banerjee et al., 2015; Banerjee, 2015a, 2015b).

Our techniques can also be combined with data on auto-antibodies and design specific strategies to increase levels of certain classes of protective auto-reactive (immunoglobulin M) IgM antibodies (Fattal et al., 2010). Our framework can also be used to find classes of T-regulatory cells that have a protective function in autoimmune diseases (Herwijnen et al., 2012). Future work will also investigate linking drug target databases with curated repositories of natural substances such as polyphenol (found in green tea) that have also been known to recruit T-regulatory cells (Wong et al., 2011). This may lead to novel natural compounds that have a protective function in autoimmune diseases. Finally, our framework can be extended to incorporate information on idyotypic networks of antibodies (antibodies that link to other antibodies) that may have a role in autoimmune diseases (Shoenfeld, 2004).

In summary, we present a computational framework for combining bioinformatics with network approaches along with a novel repository can enable us to find new targets for diseases. We show this here for autoimmune diseases. Our code is freely available from a repository (https://bitbucket.org/neelsoumya/ autoimmune_targets_pipeline). Computational techniques like these can shed light on the immune system in disease and help find novel therapeutic strategies.

Supplementary Information

A full list of all drugs in use or development and single nucleotide polymorphisms for Sjogren disorder and systemic lupus erythematosus, along with all code is available online (https://bitbucket.org/neelsoumya/autoimmune_targets_pipeline).

Acknowledgment

The author wishes to thank Joyeeta Ghose and Dr. Beverly Kloeppel for fruitful discussions.

References

Banerjee S. 2013. Scaling in the Immune System, PhD Thesis. University of New Mexico, USA

Banerjee S. 2015a. Analysis of a planetary scale scientific collaboration dataset reveals novel patterns. arXiv preprint arXiv:1509.07313

Banerjee S. 2015b. Optimal strategies for virus propagation. arXiv preprint arXiv: 1512.00844

Banerjee S, Guedj J, Ribeiro RM, Moses M, Perelson AS. 2016. Estimating biologically relevant parameters under uncertainty for experimental within-host murine West Nile virus infection. Journal of the Royal Society Interface, 13(117): 20160130

Banerjee S, Moses M. 2010. Scale invariance of immune system response rates and times: Perspectives on immune system architecture and implications for artificial immune systems. Swarm Intelligence, 4(4): 301-308

Banerjee S, Moses M. 2009. A hybrid agent based and differential equation model of body size effects on pathogen replication and immune system response. In: Lecture Notes in Computer Science: Vol. 5666. Artificial Immune Systems (Andrews PS et al., eds). 8th International Conference, ICARIS 2009. 14-18, Springer, Berlin, Germany

Banerjee S, van Hentenryck P, Cebrian M. 2015. Competitive dynamics between criminals and law enforcement explains the super-linear scaling of crime in cities. Palgrave Communications, 1: 15022

Fattal I, Shental N, Mevorach D, Anaya JM, Livneh A, et al. 2010. An antibody profile of systemic lupus erythematosus detected by antigen microarray. Immunology, 130(3): 337-343

Jesmin T, Waheed S, Al-Emran A. 2016. Investigation of common disease regulatory network for metabolic disorders: A bioinformatics approach. Network Biology, 6(1): 28-36

Koscielny G, An P, Carvalho-Silva D, Cham JA, et al. 2017. Open Targets: a platform for therapeutic target identification and validation. Nucleic Acids Research, 45(D1): 985-994

Shoenfeld Y. 2004. The idiotypic network in autoimmunity: Antibodies that bind antibodies that bind antibodies. Nature Medicine, 10(1): 17-18

Tacutu R, Budovsky A, Yanai H, et al. 2011. Immunoregulatory network and cancer-associated genes: molecular links and relevance to aging. Network Biology, 1(2): 112-120

van Herwijnen MJC, Wieten L, van der Zee R., van Kooten PJ, Wagenaar-Hilbers JP, et al. 2012. Regulatory T cells that recognize a ubiquitous stress-inducible self-antigen are long-lived suppressors of autoimmune arthritis. Proceedings of the National Academy of Sciences of the USA, 109(35): 14134-14139

Wong CP, Nguyen LP, Noh SK, Bray TM, et al. 2011. Induction of regulatory T cells by green tea polyphenol EGCG. Immunology Letters, 139(1–2): 7-13

Zhang WJ. 2016a. Network pharmacology: A further description. Network Pharmacology, 1(1): 1-14

Zhang WJ. 2016b. Network robustness: Implication, formulization and exploitation. Network Biology, 6(4): 75-85

10

Networks control: Introducing the degree of success and feasibility

Alessandro Ferrarini

Department of Evolutionary and Functional Biology, University of Parma, Via G. Saragat 4, I-43100 Parma, Italy

E-mail: sgtpm@libero.it, alessandro.ferrarini@unipr.it

Abstract

Taming ecological and biological networks is a key-issue. It could be used to: a) neutralize damages to ecological and biological networks, b) safeguard rare and endangered species, c) manage ecological systems at the least possible cost, and d) counteract the impacts of climate change. While I recently showed that ecological and biological networks can be efficaciously controlled both from inside (inside-control model) and outside (outside-control model), here I propose a solution to the choice of the most feasible solution to network control. To do this, I introduce the concepts of control success and feasibility.

Keywords edges control feasibility; control success; control uncertainty; genetic algorithms; network control; stochastic simulations.

1 Introduction

Recently, I proposed that ecological and biological networks can be controlled by coupling network dynamics and evolutionary modelling (Ferrarini, 2011). They can be efficaciously tamed from outside (Ferrarini, 2013a), but also through the use of endogenous controllers (Ferrarini, 2013b). These two approaches are different from both a theoretical and methodological viewpoint. The endogenous control requires that the network is optimized at the beginning of its dynamics (by acting upon nodes, edges or both) so that it will then go inertially to the desired state. Instead, the exogenous control requires that exogenous controllers act upon the network at each cycle. *A priori*, it's hard to say which of the two approaches is more effective, it mainly depends on the kind of ecological or biological network one is dealing with.

In another paper (Ferrarini, 2013c), I have faced a further important question: how reliable is the achieved solution? In other words, which is the degree of uncertainty about getting the desired result if values of edges and nodes were a bit different from optimized ones? This is a pivotal question, because it's not assured that while managing a certain system we are able to impose to nodes and edges exactly the optimized values we would need in order to achieve the desired results. In order to face this topic, I have coined a 3-parts framework (network dynamics - genetic optimization - stochastic simulations).

Here I propose a solution to the choice of the most feasible solution to network control. To do this, I introduce the concepts of control success and feasibility.

2 Mathematical Formulation

Most real systems' dynamics can be modelled and simulated as follows (Liu et al., 2011; Slotine and Li, 1991):

$$
\begin{cases}
\dfrac{dS_1}{dt} = a_{11}S_1 + \ldots + a_{1n}S_n + I_1 + O_1 \\
\ldots \\
\dfrac{dS_n}{dt} = a_{n1}S_1 + \ldots + a_{nn}S_n + I_n + O_n
\end{cases}
\tag{1}
$$

where S_i is the number of individuals (or the total biomass, or the covered surface in case of plant species) of the generic *i-th* species, while I and O represent inputs and outputs from/to outside.

Biological and ecological systems can be tamed from outside using the following 1-external-controller model (Ferrarini 2013a):

$$
\begin{cases}
\left(\dfrac{dS_1}{dt}\right)_{OPT} = a_{11}S_1 + \ldots + a_{1n}S_n + I_1 + O_1 + c_{11*}C_{1*} \\
\ldots \\
\left(\dfrac{dS_n}{dt}\right)_{OPT} = a_{n1}S_1 + \ldots + a_{nn}S_n + I_n + O_n + c_{n1*}C_{1*} \\
\dfrac{dC_1}{dt} = f_1 S_1 + \ldots + f_n S_n
\end{cases}
\tag{2}
$$

where asterisks stand for the genetic optimization (Holland 1975) of exogenous node's edges (i.e., coefficients of interaction with the inner system) and exogenous node's stock, i.e. the modification of such values at the beginning of network dynamics in order to get a certain goal (e.g., maximization of the final value of a certain variable). There's 1 controller C_1 that, in some cases, can also receive feedbacks from the network, It's clear that also the feedback dC_1/dt to the controller could be subject to genetic control by taming $<f_1 \ldots f_n>$.

In case 1 controller is not enough, the model in (2) must be expanded to the following *k*-external-controllers model (Ferrarini 2013a):

$$\begin{cases} \dfrac{dS_1}{dt} = a_{11}S_1 + \ldots + a_{1n}S_n + I_1 + O_1 + c_{11*}C_{1*} + \ldots + c_{1k*}C_{k*} \\[2mm] \ldots \\[2mm] \dfrac{dS_n}{dt} = a_{n1}S_1 + \ldots + a_{nn}S_n + I_n + O_n + c_{n1*}C_{1*} + \ldots + c_{nk*}C_{k*} \\[2mm] \dfrac{dC_1}{dt} = f_{11}S_1 + \ldots + f_{1n}S_n \\[2mm] \ldots \\[2mm] \dfrac{dC_k}{dt} = f_{k1}S_1 + \ldots + f_{kn}S_n \end{cases} \qquad (3)$$

Alternatively, an ecological or biological network can be controlled from inside using the following control model (Ferrarini, 2013b):

$$\begin{cases} \left(\dfrac{dS_1}{dt}\right)_{OPT} = a_{11*}S_1^* + \ldots + a_{1n*}S_n^* + I_{1*} + O_{1*} \\[2mm] \ldots \\[2mm] \left(\dfrac{dS_n}{dt}\right)_{OPT} = a_{n1*}S_1^* + \ldots + a_{nn*}S_n^* + I_{n*} + O_{n*} \end{cases} \qquad (4)$$

where asterisks stand for the optimization of edges (i.e., coefficients of interaction among variables) or nodes (i.e., initial stocks), that is the modification of their values at the beginning of the network dynamics in order to get a certain goal.

After optimization is reached, the degree of uncertainty of (2), (3) and (4) about getting the desired result can be computed as (Ferrarini, 2013c):

$$\begin{cases} \left(\dfrac{dS_1}{dt}\right)_{OPT} = \underline{a_{11*}S_1^*} + \ldots + \underline{a_{1n*}S_n^*} + \underline{I_{1*}} + \underline{O_{1*}} \\[2mm] \ldots \\[2mm] \left(\dfrac{dS_n}{dt}\right)_{OPT} = \underline{a_{n1*}S_1^*} + \ldots + \underline{a_{nn*}S_n^*} + \underline{I_{n*}} + \underline{O_{n*}} \end{cases} \qquad (5)$$

where:

$$\begin{cases} 0.95 * a_{ij*} \leq \underline{a_{ij}} \leq 1.05 * a_{ij*} \\[2mm] 0.95 * S_j^* \leq \underline{S_j} \leq 1.05 * S_j^* \end{cases} \qquad (6)$$

or alternatively:

$$\begin{cases} 0.9 * a_{ij*} \leq \underline{a_{ij}} \leq 1.1 * a_{ij*} \\ 0.9 * S_j^* \leq \underline{S_j} \leq 1.1 * S_j^* \end{cases} \tag{7}$$

Hence, $\underline{a_{ij}}$ represents a 5% (or 10%) uncertainty about a_{ij*}, while $\underline{S_j}$ represents a 5% (or 10%) uncertainty about S_j^*. If, after genetic optimization, we stochastically vary n times (e.g. 10,000 times) a_{ij*} and S_j^*, we are able to compute how many times such uncertainty makes the optimization procedure useless. Hence, uncertainty about network control can be computed as (Ferrarini, 2013c):

$$U_\% = 100 * \frac{k}{n} \tag{8}$$

where k is the number of stochastic simulations acting upon optimized parameters that make the optimization procedure useless (i.e. the goal of optimization is not reached).

Now, let's assign to each species a weight of importance σ_i:

$$\sigma_i : \begin{cases} > 0 & \text{for benefit species (or network actors)} \\ 0 & \text{for species (or network actors) of no interest} \\ <0 & \text{for cost species (or network actors)} \end{cases} \tag{9}$$

I suggest here that the degree of success DS_i of network control for each *i-th* species can be computed as the weighted difference between the optimized dynamic of the species (S_i^{opt}: how it goes, at equilibrium, after optimization) and the inertial one (S_i^{in}: how it would go, at equilibrium, without optimization):

$$DS_i = \sigma_i * \Delta_i = \sigma_i * (S_i^{opt} - S_i^{in}) \tag{10}$$

DS_i is positive if a benefit species has increased thanks to the network control or a cost species has decreased. Instead it's negative in case a benefit species has decreased due to the network control or a cost species has increased. The overall degree of success of network control for n species (or network actors) can be hence calculated as:

$$DS = \sum_{i=1}^{n} \sigma_i * \Delta_i \tag{11}$$

Now I define the degree of feasibility F of network control as:

$$F = \frac{DS}{1 + U_\%} \tag{12}$$

where the constant *1* has been added to avoid that the denominator goes to 0. As a first approximation, I suggest that the weight of importance σ_i should go from -1 to +1. But, in order to give DS and $1+U_\%$ the same order of magnitude, I suggest that σ_i should be set so that:

$$\begin{cases} 1 \le \dfrac{DS}{1+U_{\%}} \le 10 \\ \\ or \\ \\ 1 \le \dfrac{1+U_{\%}}{DS} \le 10 \end{cases} \tag{13}$$

It's clear that F is a 3D surface equation in the form:

$$Z = \frac{x}{1+y} \tag{14}$$

Hence the feasibility surface is like in Fig. 1.

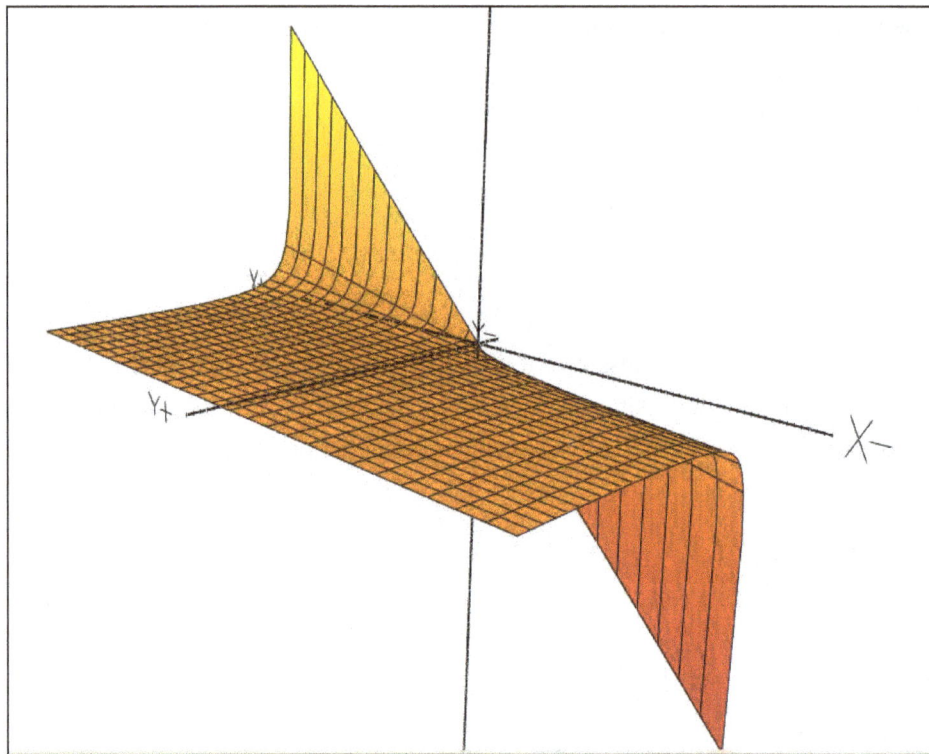

Fig. 1 Feasibility surface as a function of control success (X-axis) and control uncertainty (Y-axis).

Since many solutions to network control can be found using the previous control models (2), (3) and (4), each j-*th* solution will receive its degree of feasibility

$$F_j = \frac{DS_j}{1+U_{j\%}} \tag{15}$$

and the best solution to network control will be the one with

$$max(F_j) \tag{16}$$

3 Conclusions

Taming ecological and biological networks is a key-issue. It could be used to: a) neutralize damages to ecological and biological networks, b) safeguard rare and endangered species, c) manage ecological systems at the least possible cost, and d) counteract the impacts of climate change.

While I recently showed that ecological and biological networks can be efficaciously controlled both from inside (inside-control model) and outside (outside-control model), here I have proposed a solution to the choice of the most feasible solution to network control. To do this, I have introduced the concepts of control success and feasibility.

References

Ferrarini A. 2011. Some thoughts on the controllability of network systems. Network Biology, 1(3-4): 186-188

Ferrarini A. 2013a. Exogenous control of biological and ecological systems through evolutionary modelling. Proceedings of the International Academy of Ecology and Environmental Sciences, 3(3): 257-265

Ferrarini A. 2013b. Controlling ecological and biological networks *via* evolutionary modelling. Network Biology, 3(3): 97-105

Ferrarini A. 2013c. Computing the uncertainty associated with the control of ecological and biological systems. Computational Ecology and Software, 3(3): 74-80

Holland JH. 1975. Adaptation in Natural and Artificial Systems: An Introductory Analysis with Applications to Biology, Control and Artificial Intelligence. University of Michigan Press, Ann Arbor, USA

Liu YY, Slotine JJ, Barabasi AL. 2011. Controllability of complex networks. Nature, 473:167-173

Slotine JJ, Li W. 1991. Applied Nonlinear Control. Prentice-Hall, USA

11

Organizational theory: With its applications in biology and ecology

Yue Zhao[1], WenJun Zhang[1,2]
[1]Sun Yat-sen University, Guangzhou 510275, China
[2]International Academy of Ecology and Environmental Sciences, Hong Kong
E-mail: hwason@126.com,zhwj@mail.sysu.edu.cn,wjzhang@iaees.org

Abstract

Organizations are goal-directed entities which have been designed as deliberately structured and coordinated dynamic systems that connect with the external environment. Organizational theory is the study of structure, function and design of organization. It aims to solve practical problems, maximize production efficiency and make organization better function and develop. Organizational theory contains various aspects. The history, development, and thoughts of organizational theory and its applications in biology and ecology were described in present paper. We held that more studies should be conducted to apply organizational theory in natural sciences as biology and ecology.

Keywords organization; organizational theory; biology; ecology.

1 Definition

During the past hundreds of years, the definition of organization has being continuously refreshed and improved. Daft and Armstrong (2007) described organizations as goal-directed social entities which were designed as deliberately structured and coordinated dynamic systems that connect with the external environment. Tompkins (2005) held that organizational theory was the study of how and why those complicated organizations behaved as the way they were. Apparently, a complex organization is too enormous and structurally differentiated to be effectively represented by a single individual. Organizational theory is neither a single piece of theory nor an integrated body of information but a field of studies which cover various scientific disciplines and subjects. The depth and breadth of this study field is challenging numerous researchers.

Organizational design is an important field in organizational theory. The importance of improving our knowledge of it remains high in the coming future due to a series of trends, such as advances in information technology that encourage experimentation with new organizational designs; large economies like India and China are attempting to rapidly transform the organizational infrastructure of their administration; the professionalization of the NGO and charity sectors, and multinational corporations' increasing attempts to exploit globally distributed intellectual resources, etc (Puranam, 2012).

2 History of Organizational Theory

As an advanced and broad discipline, organizational theory has a very long history with a mission for pursuing scientism, managerialism and enhanced efficiency and effectiveness (Üsdiken and Leblebici, 2001). It is an ancient but also modern scientific discipline. Organization research gained its status in science since Aristotle (Rosvall and Bergstrom, 2011). The research on organizations began its journey along the civilization in the human world. However, organizational theory wasn't recognized as a scientific discipline until the 1960s (Cunliffe, 2008). Since 19th century, in particular industrial revolution, organization studies have quickly developed especially in such areas as socio-political questions (Wolin, 1961).

Cunliffe (2008) divided the developmental period of organizational theory into four stages, (1) classical and scientific management/modernism, (2) systems and contingency theories, (3) social construction, and (4) postmodernism. The first stage is classical and scientific management stage, and Adam Smith, Carl Marx, Taylor, Weber, et al., were representative researchers during this period. These researchers have drawn and distilled theories from routine and social activities, and built fundamental concepts of organization. The second stage, system and contingency theories, i.e. modernism, was governed by such researchers as Parsons, Woodward and others. They emphasized the optimization of production efficiency and stressed the need of treating organization as a sophisticated system (Barzilai, 2010). The third stage, social construction, was mainly represented by Berger, Goffman, Weick, etc. They considered that the sharement between organizations was important since they were actually communities which interacted with each other. In the last stage, postmodernism, more researchers appeared, such as Harvey, Cooper and Burrell. During this period, various thoughts on organizations formed and evolved.

A little different from the classification above, Docherty (2001) classified the developmental process of organizational theory into three big stages, (1) classical theory, (2) neoclassical theory, and (3) contemporary theory. In the first stage, the mass production facilitated the overall formation of organization and relative theories. Focuses in this stage were the studies of some contents about laborers, division and scientific management, such as hierarchy, span of control, the degree of centralization and the specialization of work. Unlike the first one, in the second stage, neoclassical theory, organizational theory put its main focus upon the individuals and their mutual relationship (or interactions). In the last stage, contemporary theory, various theories appeared and organizational theory stepped into a new time.

3 Major Theories

Organizational theory came up with different theories. Here we make a summary of these theories.

3.1 Classical organizational theory

This mainly consists of three sub-theories: scientific management theory, Weber's bureaucratic theory, and administrative theory.

(a) Scientific management theory. Frederick W. Taylor was recognized as the pioneer of scientific management. He started the theory with the observation of working process and concluded that how to minimize the input, maximize the efficiency and achieve specialization and standardization. However, this theory was criticized by laborers for the reason that this system overlooked the human's perception and senses.

(b) Weber's bureaucratic theory. Acclaimed as the father of sociology, Max Weber described the basic information about bureaucratic theory. Under his bureaucracy's condition, an organization is governed by top-down rules and regulations; employees work on strictly defined responsibility and own restrained power. He also described the essential means to draw a picture about organizational theory based on the historical point.

(c) Administrative theory. The major representative of administrative theory was H. Fayol. This theory takes the form of hierarchical pyramid as its structure. He developed fourteen principles to advise managers to mandate and fulfill their responsibility. In addition, he outlined five basic elements of management: planning, organizing, command, coordination and control.

3.2 Neoclassical organizational theory

This theory was led by the studies of Hawthorne in the 1920s. Significantly different from the early thoughts and approaches, this theory particularly emphasized the importance of personnel relationships among the workers, employees and managers, reflecting the growing need of humane and emotional care of workers. Laborers with high concentration and volition contribute positively and meticulously, so the company and factory would benefit more and function better. A lot of studies were thus conducted by sociologists and psychologists, among which Elton Mayo contributed the most.

3.3 Contemporary theories

They are also called modern theories. Dozens of new theories have appeared in modern times. Modern theories evolved quickly in rapidly-changing environments into various shapes and structures. Several major theories or methods are described as follows.

(a) The system method. According to this method, an organization is viewed as a system which composes of many mutually connected components, aimed to obtain benefits both internally and externally. Overall an organization consists of three parts: components, linking processes, and goals of organization (Bakke, 1959).

(b) Contingency theory. Contingency theory, proposed by Lawrence and Lorsch, suggests that there is no best way to direct an enterprise, because the requirements for corporations vary enormously in different environments and conditions. A series of factors may work as variables, including environmental uncertainty, technology, size, strategy, resource dependence and public accountability (Tompkins, 2005).

(c) Other theories. There are some other theories that play a better role in the modern organizational theory, such as quality management theory, organizational culture, leadership theory, and so forth.

4 Further Explanation of Organizational Theory

Donaldson (2003) described organizational theory as a positive science. Driven by the environment, scientific methods validated and testified these positive but normative theories. Up to date, organizational science has made huge progress by using the positivist approach. Organizational theory has proved strong potentiality in the future as regards being pursued positively.

Hatch and Yanow (2003) called organizational theory an interpretive science. Many interpretive researchers who like them strongly held that social world and natural world ought to be ascertained in different ways.

Willmott (2003) viewed organizational theory a critical science. While Chia (2003) thought organizational theory as a postmodern science and draw our attention to the requirement for managers and policymakers. Obviously they were more aware of the basic information and situation of our society and industries.

An organization cannot thrive without successful and powerful traits. Faced with threats and chances, it should be sensitive to external changes and keeping adapting and learning (Hannah and Lester, 2009). Learning is not limited to the scope of knowledge per se but a "problem-oriented action" or "knowing" (Kuhn and Jackson, 2008). Roberts (2007) probed into the knowledge in the contemporary organization by summarizing several influential books and has managed to address relevant problems. Rashman et al. (2009) reviewed the literature on organizational learning and knowledge relevant with public organizations particularly, and maintained their uniqueness by using the dynamic model. The external situations in the environment, are also vital issues. Analyzing the community context will revitalize the research on organizations (Freeman and Audia, 2006), since organizations function with other social units interdependently. King et al. (2010) noted that we should locate the organization in a wider social landscape and then explore its uniqueness as a social actor.

In addition to external conditions, internal components are important. An organization cannot survive or exist without rational structure and design of the system. Rank (2008) argued that although considerable researches aimed at unveiling the complicated function of organizational systems, little attention has been given to the "structural interdependencies between formal organizations and informal networks".

Ocassionally, some peculiar things could boost our understanding on organizational theory. Jones and Munro (2005) examined the works of eighteen researchers on modern organizational theory in the last twenty years. Many topics and debates were discussed including some basic concepts and postmodernism. Warner (2007) mentioned modern literary guru- Franz Kafka. His works shed light on the deep examining of organizations, and further being compared with Max Weber. In addition, some researchers explored the organizational theory in terms of its logics with novel insights and methods (Hannan, 2007; Kamps, 2009; Durand, 2008), which was mainly shaped in the book, Logics of Organizational Theory: Audiences, Codes, and Ecologies (Hannan, 2007). But it has focused to the entire process of theory-shaping, deviating from the traditional way of organizational ecology (Kamps, 2009). Santos and Eisenhardt (2005) stressed organizational boundaries, which may facilitate the understanding of organizations. Kulic and Baker (2008) also held that the boundaries were hard to be drawn clearly under real-world situations. As a response, they proposed another method to cover various views of organizations in a simulative environment using computational organizational theory. Audia et al. (2006) connected the theories of organizational ecology and social network and dug into the variations "in rates of foundings over geographic locales" affected by the structure of relations in various populations.

5 Organizology: A Proposed New Science on Organization

It is true that one cannot find any entry about this terminology in Webster's dictionary and even by searching it on Internet. We could only get just over six hundred outcomes. Actually, this terminology was crafted by Aleinikov in 2003. After sharing the Sedov's statement that moving matter has only two characteristics: the intensity of movement and the organization of movement, Aleinikov proposed organizology- the science of organization of movement, following this binary logic (Aleinikov and Smarsh, 2010).

Science per se needs to be ordered and organized properly. Moreover, a science of organization should be founded considering the development of science and organization. But this did not happen until the recent years owing to the absent of measurement or organization. The proposition of organizology is a beneficial attempt in this aspect. Organizology was founded on one basic measurement to address complex problems with one basic explanation (Aleinikov, 2004). But the defining and refining of this measurement proved to be a challenge and a new unit of organization expressed by the formula T/L (time divided by space) was offered. Also, a few cases were provided to explain its nature in different kinds of organizations. And this unit of organization was named "aleandr" (Aleinikov, 2005). Furthermore, an accurate prediction based on mathematics was given to elucidate this science (Aleinikov and Gera, 2006).

It could be expected that any nascent theory would confront controversy and criticism. Organizology is no exception. Anyway, a breakthrough thought to the classic theory ought to be encouraged.

6 Future Prospect of Organizational Theory

Organizational theory is an unavoidable derivative of historical development, under the impact of multiple forces: industrial revolutions, technological revolution, digital revolution and the third industrial revolution (Rifkin, 2013). A series of revolutions have produced novel thoughts and minds, industries and sciences, craftsmanship and technologies, which endowed humans with new lifestyles and jobs, new ideas to understand the world and new managerial methods. Since the late 1980s, the trajectory of organizational theory has changed from "paradigm-driven work to problem-driven work" (Davis and Marquis, 2005). Huge changes of organization theory come as the result of discontinuous and fast-changing environment (Marshak, 2004). Accordingly, a new model of reconfigurable organization has been given (Stefanovic et al., 2011). Organizational theory takes nutrients from other scientific fields and industries unprecedentedly, and eventually evolves into various shapes and colors. No one can give an exact answer to this question: what is the future of organizational theory? Every research manages to elucidate his own ideas and imaginations and as a consequence, various thoughts emerge and evolve.

Walsh et al. (2006) poses three fundamental and difficult questions about the future of organizational theory. "How can we understand, live in and live with today's organizations?" He held that the trend is hardly to be traced and predicted, considering this fast-changing world and the influences of globalization. But as difficult as it may be, we could work hard to collect data and empirical evidences and get a handle of it. A little different, Burrell (2003) proposed another two questions concerning the subsequent research areas to deal with and methodology and epistemology approaches to use. Davis and Marquis (2005) argued that the central target in the twenty-first century was to explain the economic institutions.

7 Applications of Organizational Theory in Biology and Ecology

Phillips (1992) considered that some methods used in the natural science had been partly accepted by social science and vice versa (Burrell and Morgan, 1979). Biological systems like social systems represent hierarchical organizations with sub-modules that cover multiple scales (Rosvall and Bergstrom, 2011). Besides contingency theory, which is basically an organic analogy, organizational ecology has been an attractive research in the last decade mainly in the USA. It aims to explain "how social, economic and political conditions affect the relative abundance and diversity of organizations and to account for their changing composition over time" (Baum and Amburgey, 2005), and to emphasize the "evolutionary dynamics of processes" behind them (Singh and Lumsden, 1990). To understand the organizational diversity means to answer a question: "Why are there so many (or so few) kinds of organizations?" (Hannan and Freeman, 1993). This theory uses biological and ecological models to analyze businesses and issues about organizations (Clegg and Hardy, 1999). Dobrev et al. (2006) argued this theory as a "research paradigm", and by this theory, "multivariate models" are used for various potential reasons. Hannan and Freeman (1977) proposed a perspective of population ecology to analyze the relationship between organization and environment. Carroll (1984) reviewed some research on organizational ecology and especially distinguished three different levels of analysis and methods: organizational, population and community, followed by developmental, selection and macro-evolutionary method, respectively. And a development on organizational taxonomies was also recommended. Within more than thirty years, organizational ecology has taken a long step, but it cannot sleep on the pillow of past merits and achievements. Innovations of knowledge and theory are required to revive it. Particularly, ecological theory is of vital importance and most connected with evolution and this is doomed to be given more attention (Amburgey and Rao, 1996).

Baum and Amburgey (2005) maintained that ecological approaches were radically different from the traditional ones. The former methods focus on the contextual factors. However organizational ecology that has built mainly on the population models indeed confuses some sociologists (Hannan, 2005). Using biological theories or metaphors to explain organizational changes has been critically misunderstood (Singh and Lumsden, 1990).

Reuter et al. (2010) proposed new approaches to explain ecological interactions across scales. They stressed multiple organizational hierarchies and their mutual effects. Cross-scale interactions are among the most prominent concerns in ecology and biodiversity problems, invasive species and long-term effects of habitat change (Kerr et al, 2007). Lidicker (2007) came up with the fourth level—"ecospace" of levels of ecological organizations, apart from the other three, organism, population and community, given the hierarchical arrangements being increasingly favored by many ecologists.

Actually, organization assembles organisms to some extent (Morgan, 2006). They function as a whole. Accordingly, organizational theory may be rationally used to fit with the natural world. Organizations interact with communities they dwell in (Freeman and Audia, 2006). The population ecologists hold the view that the ability of seizing a resource niche and defeating its rivals are really matter. The concept of ecological niche has been successfully adopted by organizational theory researchers. In contrast with the niche in natural world, the organizational niche reveals itself in social and economic world (Boone et al., 2002). Therefore, sociologists have made their minds to discover the appropriate niche by which an organization can develop and prosper (Hannan, 2005). In fact, this view does not satisfy many organization theorists, since they emphasize the role of managers and decision makers but not merely the viewpoint that environments choose organizations (Morgan, 2006). For instance, some critics refuted contingency theory that an organization can be self-adjusted to fully utilize its environment (Pfeffer and Salancik, 2003).

Great theories pave the way for the dawn of typical practices. Couzin (2006) has paid his attention to the social organization in one of the most complex "fission-fusion" systems in nature- elephant populations. However fully elucidating and understanding this system by using various technologies and constructing mathematical models proves a huge challenge still. Co-evolutionary method plays an important role in helping researchers who study the natural environment and organizations to revise their organizational theory (Porter, 2006). It works as a propeller (Lewin and Volberda, 2003). But it should be noted that organizational co-evolution involves some aspects absent from its biological counterpart, knowledge, learning, demand, actor traits and behavior, strategy and tactics (Malerba, 2006; Zhang, 2012). Organizations not only hold a position in large-scale environments but also make them the fundamental concepts in cell biology: self-organization emerged to explain and understand the components and compartments of the cells (Karsenti, 2008).

Organizational theory is evolving rapidly, melting with other disciplines, such as ecology. Examining the history and prospecting the future can be helpful for us to understand and develop organizational theory. So far, organizational theory is mainly a discipline of social sciences. We suggest that more studies should be conducted to apply organizational theory in natural sciences as biology (cell biology, network biology, etc.) and ecology.

References

Aleinikov AG, Smarsh DA. 2010. Law of conservation of intensivity. Proceedings of the Academy of Strategic Management, 9(1): 3

Aleinikov AG, Gera R. 2006. Mathematical predictions of organizology, the new science of organization. Proceedings of the Fall 2006 International Conference, Reno, Nevada, USA, 11-12

Aleinikov AG. 2004. Organizology, new science of organization: A breakthrough in scientific explanation of the world. Proceedings of the Allied Academies Spring International Conference, 1-2

Aleinikov AG. 2005. Organizology: The main unit of measurement for the new science. Proceedings of the Academy of Creativity and Innovation, 2(1): 15

Amburgey TL, Rao H. 1996. Organizational ecology: past, present, and future directions. Academy of Management Journal, 39(5): 1265-1286

Audia P, Bielefeld W, Dowell M. 2006. Organizational foundings in community context: instruments manufacturers and their interrelationship with other organization. Administrative Science Quarterly, 51(3): 381-419

Bakke WE. 1959. Concept of the social organization. In: Modern organization theory (Haire M, ed). Chapman and Hall, USA

Barzilai K. 2010. Organizational theory. http://www.cwru.edu/med/epidbio/mphp439/Organizational_Theory.htm. Retrieved August 20, 2010

Baum JAC, Amburgey TL. 2005. Organizational ecology. In: The Blackwell Companion to Organizations (Baum JAC, ed). Wiley, USA

Boone C, van Witteloostuijn A, Carroll GR. 2002. Resource distributions and market partitioning: Dutch daily newspapers, 1968 to 1994. American Sociological Review, 67(3): 408-431

Burrell G, Morgan G. 1979. Sociological Paradigms and Organizational Analysis. Ashgate Publishing Limited, USA

Carroll GR. 1984. Organizational ecology. Annual Reviews of Sociology, 10: 71-98

Chia R. 2003. Organizational theory as a postmodern science. In: The Oxford Handbook of Organizational Theory (Tsoukas H, Knudsen C, eds). Oxford University Press, UK

Clegg SR, Hardy C. 1999. Studying Organization. SAGE Publication Ltd, UK

Couzin LD. 2006. Behavioral ecology: social organization in fission-fusion societies. Current Biology, 16(5): 170

Cunliffe AL. 2008. Organizational Theory. SAGE Publications Inc., USA

Daft RL, Armstrong A. 2007. Organizational Theory and Design. Graphic World Inc., USA

Davis GF, Marquis C. 2005. Prospects for organizational theory in the early twenty-first century: institutional fields and mechanisms. Organization Science, 16(4): 332-343

Dobrev SD, van Witteloostuijn A, Baum JAC. 2006. Introduction: ecology versus strategy or strategy and ecology? Advances in Strategic Management, 23: 1-26

Docherty JP. 2001. Organizational theory. In: Textbook of Administrative Psychiatry: New Concepts for a Changing Behavioral Health System (Talbott JA, Hales RE, eds). American Psychiatric Publishing Inc., USA

Donaldson L. 2003. Organizational theory as a positive science. In: The Oxford Handbook of Organizational Theory (Tsoukas H, Knudsen C, eds). Oxford University Press, UK

Durand R. 2008. Logics of organizational theory: audiences, codes, and ecologies. Contemporary Sociology, 37(6): 605-606

Freeman JH, Audia PG. 2006. Community ecology and the sociology of organizations. Annual Review of Sociology, 32: 145-169

Hannan MT. 2005. Ecologies of organizations: diversity and identity. Journal of Economic Perspectives, 19(1): 51-70

Hannan MT. 2007. Logics of Organizational Theory: Audiences, Codes, and Ecologies. Princeton University Press, USA

Hannan MT, Freeman J. 1977. The population ecology of organizations. American Journal of Sociology, 82(5): 929-964

Hannan MT, Freeman J. 1993. Organizational Ecology. Harvard University Press, USA

Hannah ST, Lester PB. A multilevel approach to building and leading learning organizations. The Leadership Quarterly, 20: 34-48

Hatch MJ, Yanow D. 2003. Organizational theory as an interpretive science. In: The Oxford Handbook of Organizational Theory (Tsoukas H, Knudsen C, eds). Oxford University Press, UK

Jones C, Munro R. 2005. Organizational theory, 1985-2005. Sociological Review, 53(1): 1-15

Kamps J. 2009. Logics of organizational theory: audiences, codes, and ecologies. Administrative Science

Quarterly, 54(2): 350-353

Karsenti E. 2008. Self-organization in cell biology: a brief history. Natural reviews molecular cell biology, 9(3): 255-262

Kerr JT, Kharouba HM, Currie DJ. 2007. The macroecological contribution to global change solutions. Science, 316: 1581-1584

King BG, Felin T, Whetten DA. 2010. Finding the organization in organizational theory: a meta-theory of the organization as a social actor. Organization Science, 21(1): 290-305

Kuhn T, Jackson MH. 2008. Accomplishing knowledge: a framework for investigating knowing in organizations. Communication Quarterly, 21(4): 454-485

Kulik BW, Baker T. 2008. Putting the organization back into computational organizational theory: a complex Perrowian model of organizational action. Computational and Mathematical Organizational Theory, 14(2): 84-119

Lewin A, Volberda H. 2003. The future of organization studies: Beyond the selection – adaptation debate. In: The Oxford Handbook of Organizational Theory (Tsoukas H, Knudsen C, eds). Oxford University Press, UK, 568-595

Lidicker WZ Jr. 2007. Levels of organizations in biology: on the nature and nomenclature of ecology's fourth level. Biological Review, 83: 71-78

Malerba F. 2006. Innovation and the evolution of industries. Journal of Evolutionary Economics, 16: 3-23

Marshak RJ. 2004. Morphing: the leading edge of organizational change in the twenty-first century. Organization Development Journal, 22(3): 8

Morgan G. 2006. Images of Organization. Sage Publications Inc., UK

Pfeffer J, Salancik GR. 2003. The External Control of Organizations: A Resource Dependence Perspective. Harper and Row, USA

Phillips DC. 1992. The Social Scientist's Bestiary: A Guide to Fabled Threats to, and Defenses of, Naturalistic Social Science. Pergammon Press, Oxford, UK

Porter TB. 2006. Coevolution as a research framework for organizations and the natural environment. Organization and Environment, 19(4): 1-26

Puranam P. 2012. A future for the science of organization design. Journal of Organization Design, 1(1): 18-19

Rank ON. 2008. Formal structures and informal networks: structural analysis in organizations. Scandinavian Journal of Management, 24: 145-161

Rashman L, Withers E, Hartley J. 2009. Organizational learning and knowledge in public service organizations: a systematic review of the literature. International Journal of Management Reviews, 11(4): 463-494

Reuter H, Jopp F, Munkemuller, T, et al. 2010. Ecological hierarchies and self-organization—pattern analysis, modeling and process integration across scales. Basic and Applied Ecology, 11: 572-581

Rifkin J. 2013. The Third Industrial Revolution: How Lateral Power Is Transforming Energy, The Economy, and The World. Palgrave Macmillan, USA

Roberts J. 2007. Knowledge in the organization of contemporary business and economy. Journal of Management Studies, 44(4): 656-668

Rosvall M, Bergstrom CT. 2011. Multilevel compression of random walks on networks reveals hierarchical organization in large integrated systems. Plos one, 6(4): e18209

Santos FM, Eisenhardt KM. 2005. Organizational boundaries and theories of organization. Organization Science, 16(5): 491-508

Singh JV, Lumsden CJ. 1990. Theory and research in organizational ecology. Annual Review of Sociology, 16: 161-195

Stefanovic I, Prokić S, Vukosavljević D. 2011. The response to the changing landscape of tomorrow: reconfigurable organizations. African Journal of Business Management, 5(35): 13344-13351

Tompkins JR. 2005. Organizational Theory and Public Management. Clark Baxter Press, USA

Üsdiken B, Leblebici H. 2001. Organizational theory. In: Handbook of Industrial, Work and Organizational Psychology. Sage Publication Press, UK

Walsh JP, Meyer, AD, Schoonhoven CB. 2006. A future for organizational theory: living in and living with changing organizations. Organization Science, 17(5): 657-671

Warner M. 2007. Kafka, Weber and organizational theory. Human Relations, 60(7): 1019-1038

Willmott H. 2003. Organizational Theory as a critical science? In: The Oxford handbook of Organizational Theory (Tsoukas H, Knudsen C, eds). Oxford University Press, UK

Wolin S. 1961. Politics and Vision. Allen and Unwin, UK

Zhang WJ. 2012. Computational Ecology: Graphs, Networks and Agent-based Modeling. World Scientific, Singapore

12

A node degree dependent random perturbation method for prediction of missing links in the network

WenJun Zhang

School of Life Sciences, Sun Yat-sen University, Guangzhou 510275, China; International Academy of Ecology and Environmental Sciences, Hong Kong

E-mail: zhwj@mail.sysu.edu.cn, wjzhang@iaees.org

Abstract

In present study, I proposed a node degree dependent random perturbation algorithm for prediction of missing links in the network. In the algorithm, I assume that a node with more existing links harbors more missing links. There are two rules. Rule 1 means that a randomly chosen node tends to connect to the node with greater degree. Rule 2 means that a link tends to be created between two nodes with greater degrees. Missing links of some tumor related networks (pathways) are predicted. The results prove that the prediction efficiency and percentage of correctly predicted links against predicted missing links with the algorithm increases as the increase of network complexity. The required number for finding true missing links in the predicted list reduces as the increase of network complexity. Prediction efficiency is complexity-depedent only. Matlab codes of the algorithm are given also. Finally, prospect of prediction for missing links is briefly reviewed. So far all prediction methods based on static topological structure only (represented by adjacency matrix) seems to be low efficient. Network evolution based, node similarity based, and sampling based (correlation based) methods are expected to be the most promising in the future.

Keywords missing links; network; rules; node degree; random perturbation; prediction; likelihood.

1 Introduction

Many biological networks (food webs, protein–protein interaction networks and metabolic networks, etc) are incomplete networks due to missing links. For example, 80% of the molecular interactions in cells of Yeast (Yu et al., 2008) and 99.7% interactions of human (Amaral, 2008) are unknown. An incomplete network occurs due to our limited knowledge on the network, or the network is in evolution and thus more links or even nodes are expected with time. Link (connection) prediction tries to estimate the likelihood of the existence of a link between two nodes based on observed links and (or) the attributes of nodes (Zhang, 2015d; Zhou, 2015). Link prediction can largely reduce the experimental costs for link finding. Also, link finding algorithms can be used to predict the links that may appear in the future of evolving networks (Lü and Zhou, 2011; Lü et al.,

2012; Zhou, 2015). So far, numerous research on link prediction have been conducted (Clauset et al., 2008; Guimera and Sales-Pardo, 2009; Barzel and Barabási, 2013; Bastiaens et al., 2015; Lü et al., 2015; Zhang, 2015b, 2015c, 2015d, 2016b; Zhang and Li, 2015; Zhao et al., 2015; Zhou, 2015). In present study, I will propose an algorithm for prediction of missing links in the network, in which the likelihood of missing links of a node depends on the node degree.

2 Methods

2.1 Algorithm

Link prediction is closely correlated with network evolution. Following the principle of network evolution of Zhang's model (Zhang, 2016a), in present algorithm I assume that a node with more existing links harbors more missing links. It is a reasonable and practical assumption because new nodes tend to connect the nodes with more links (Barabasi and Albert, 1999; Zhang, 2012a; Zhang, 2016a).

Assume there are totally v nodes in the network being predicted, and adjacency matrix of the network is $d=(d_{ij})$, $i,j=1,2,...,v$, where $d_{ij}=d_{ji}$, $d_{ii}=0$, and if $d_{ij}=1$ or $d_{ji}=1$, there is a link (connection) between nodes i and j. The adjacency matrix of the network for missing links only is $D=(D_{ij})$, $i,j=1,2,...,v$. The procedures are as follows

(1) Calculate the expected missing links to be predicted, $m=m'\times per$, where m' is the total links of the network, per is the perturbation rate, and $per=0.2, 0.3$, etc., which represents a percentage increment of links in the network perturbation.

(2) Calculate the degree of node, $a_i(t)$, $i=1,2,...,v$. The cumulative attraction strength of node 1 to node i is

$$p_i(t) = \sum_{j=1}^{i} a_j(t)^{\lambda(t,a_j)} / \sum_{j=1}^{v} a_j(t)^{\lambda(t,a_j)}$$

where λ is attraction factor, $\lambda>0$. For example, $\lambda=1.2, 1.5$, etc.

(3) Generate missing links. Let $p_0=0$, and generate two random values w and u. For $p_0, p_1, p_2,..., p_v$, one of the following two rules is used

Rule 1: if $(j-1)/v \leq w \leq j/v$, $p_{k-1} \leq u \leq p_k$, $k \neq j$, and $d_{kj}=d_{jk}=0$, let $D_{kj}=1$ and $D_{jk}=1$, i.e., there is a missing link between nodes k and j.

Rule 2: if $p_{j-1}(t) \leq w \leq p_j(t)$, $p_{k-1}(t) \leq u \leq p_k(t)$, $k \neq j$, and $d_{kj}=d_{jk}=0$, let $D_{kj}=1$ and $D_{jk}=1$, i.e., there is a missing link between nodes k and j.

Rule 1 means that a randomly chosen node tends to connect to the node with greater degree. Rule 2 means that a link tends to be created between two nodes with greater degrees. By doing so, a new link is found. Repeat the procedure m times to produce m (missing) links. By doing so, an adjacency matrix of the network for missing links only, $D=(D_{ij})$, $i,j=1,2,...,v$, is generated.

(4) Return (3) to perform the next prediction, until the desired simulation times are achieved.

(5) Calculate mean number (likelihood) of predicted missing links, and rank the likelihood from greater to smaller. The first m links are the predicted missing links with maximal likelihood.

The following are Matlab codes of the algorithm (linksPrediction.m)

```
%Reference: Zhang WJ. 2016. A node degree dependent random perturbation method for prediction of missing links in the
network. Network Biology, 6(1): 1-11
clear
choice=input('Input the type (1 or 2) of data file of the network from which missing links are ready to be predicted (1: adjacency
matrix; 2: two array): ');
```

```
disp('Adjacency matrix: d=(dij)m*m, where m is the number of nodes in the network. dij=1, if vi and vj are adjacent, and dij=0, if vi and vj are not adjacent; i, j=1,2,…, m');
disp('Two array: there are two columns, A1 and A2, in the data file; an element of A1 stores a node of a link and the corresponding element of A2 stores another node of the link. ');
if (choice==1)
adjstr=input('Input the file name of adjacency matrix from which missing links are ready to be predicted (e.g., raw.txt, raw.xls, etc. Adjacency matrix is d=(dij)m*m, where m is the number of nodes in the network. dij=1, if vi and vj are adjacent, and dij=0, if vi and vj are not adjacent; i, j=1,2,…, m: ','s');
end
if (choice==2)
adjstr=input('Input the file name of two array of the network from which missing links are ready to be predicted (e.g., raw.txt, raw.xls, etc. There are two columns, A1 and A2, in the data file; an element of A1 stores a node of a link and the corresponding element of A2 stores another node of the link: ','s');
end
rule=input('Input the rule type (1 or 2) used in the algorithm: ');
pro=input('Input perturbation rate to increase missing links of the network (e.g, 0.2, 0.3, etc.): ');
lamda=input('Attraction factor of nodes (lamda>0; e.g., 1.3, 1.5, etc.)= ');
simu=input('Input the simulation times (e.g, 100, 200, etc.): ');
if (choice==1) adjmat=load(adjstr); v=size(adjmat,2); end
if (choice==2)
twoarray=load(adjstr);
nn=size(twoarray,1);
v=max(max(twoarray));
for i=1:nn
adjmat(twoarray(i,1),twoarray(i,2))=1;
adjmat(twoarray(i,2),twoarray(i,1))=1;
end; end
degr=sum(adjmat);
m=round(sum(degr)/2*pro);
fprintf('\nAdjacency matrix of the original network\n')
disp([adjmat])
fprintf('\nNode degrees of adjacency matrix of the original network\n')
disp([degr])
fprintf(['\nMean of node degrees of the original network: ' num2str(mean(degr)) '\n\n'])
cnow=(sum(degr)/2)/((v^2-v)/2);
fprintf(['\nConnectance=' num2str(cnow) '\n'])
summ=sum(degr);
summa=sum(degr.*(degr-1));
h=v*summa/(summ*(summ-1));
fprintf(['\nAggregation index (AI) of node degrees=' num2str(h) '\n'])
cv=(std(degr))^2/mean(degr);
fprintf(['\nCoefficient of variation (CV) of node degrees=' num2str(cv) '\n'])
summ=v*(v-1)/2;
su=zeros(summ,2*simu);
prop=zeros(1,v);
```

```
proptot=zeros(v);
degrr=degr.^lamda;
prop(1)=degrr(1)/sum(degrr);
for i=2:v;
prop(i)=prop(i-1)+degrr(i)/sum(degrr);
end
for siml=1:simu
adj=zeros(v);
temp=zeros(m,2);
mm=1;
while (v>0)
rep=0;
while (v>0)
propp=prop;
if ((rep==0) & (rule==1))
for i=1:v;
propp(i)=i/v;
end; end
ran=rand();
for j=1:v
if (j==1) st=0; end
if (j>=2) st=propp(j-1); end
if ((ran>=st) & (ran<propp(j))) rep=rep+1; id(rep)=j; break; end
end
if ((rep>=2) & (id(rep)~=id(1)))
tab=0;
for i=1:mm
if (((id(1)==temp(i,1)) & (id(rep)==temp(i,2))) | ((id(rep)==temp(i,1)) & (id(1)==temp(i,2)))) tab=1; break; end
end
if (tab==1) continue; end;
temp(mm,1)=id(1); temp(mm,2)=id(rep);
break;
end; end
if (adjmat(id(1),id(rep))==0) adj(id(1),id(rep))=1; adj(id(rep),id(1))=1; mm=mm+1; end;
if (mm==m+1) break; end;
end
fprintf(['Simulation ' num2str(siml)])
fprintf('\n\nAdjacency matrix for predicted links only\n')
disp([adj])
[pairx,pairy]=find(adj);
temp1=pairx; temp2=pairy;
pairxs=pairx(temp1<temp2);
pairys=pairy(temp1<temp2);
ConnectionPairs=[pairxs pairys];
dm=size(ConnectionPairs,1);
```

```
su(:,siml*2-1)=[pairxs;zeros(summ-dm,1)]; su(:,siml*2)=[pairys;zeros(summ-dm,1)];
disp('Predicted links')
disp([ConnectionPairs])
end

disp('-----------------------------Summary--------------------------------')
disp(['There are totally ' num2str(sum(degr)/2) ' links in the original network'])
disp(['You wish to predict ' num2str(m) ' missing links in the original network'])
fprintf('\n');
proptot=zeros(v);
for i=1:v-1
for j=i+1:v
for k=1:simu
for l=1:v*(v-1)/2
if ((su(l,k*2-1)==i) & (su(l,k*2)==j)) proptot(i,j)=proptot(i,j)+1; proptot(j,i)=proptot(i,j); break; end
end; end; end; end
disp('Likelihood (mean number) of predicted links: ')
disp('    Node        Node        Likelihood')
s=0;
for j=1:v
for i=1:v
if (proptot(i,j)~=0) s=s+1;pairvalue(s)=proptot(i,j)/simu; end;
end; end
[pairx,pairy]=find(proptot);

result=[pairx pairy pairvalue'];
results(1,1)=result(1,1); results(1,2)=result(1,2); results(1,3)=result(1,3);
su=1;
for i=2:s
lab=0;
for j=1:i-1
if ((result(j,2)==result(i,1)) & (result(j,1)==result(i,2))) lab=1; break; end;
end
if (lab==0) su=su+1;results(su,1)=result(i,1); results(su,2)=result(i,2); results(su,3)=result(i,3); end
end
ires=sortrows(results,-3);
disp([ires])
```

2.2 Validation

In present study, I used the data of tumor related networks (pathways) (ABCAM, 2012; Huang and Zhang, 2012; Li and Zhang, 2013; Pathway Central, 2012; See supplementary material for adjacency matrices). These networks are complete. For each network, some links are removed following reverse process of the algorithm above and then predicted. The simulation times are set to be 100. The perturbation rate is *per*=~0.25. Attraction factor λ=1.5.

3 Results

3.1 Rule 1

Some of the summarized results for link prediction of tumor related networks (the pathways Ras, p53, Akt, HGF, JNK, PPAR, TGF-β, and TNF) are listed in Table 1 and 2, and the percentages of correctly predicted links with randomization method are given also. Here, the percentage of correctly predicted links against number of missing links (%) = correctly predicted links / number of missing links ×100, and the percentage of correctly predicted links against predicted missing links (%) = correctly predicted links / total of predicted missing links ×100, connectance = number of observed links / number of possible maximum number of links.

Table 1 Link prediction of Ras, p53, and Akt networks with Rule 1 (per=~0.25, λ=1.5, 100 simulations). The listed links are true links missed in the data used for predicting.

Ras				p53				Akt			
Rank	Node	Node	Likelihood	Rank	Node	Node	Likelihood	Rank	Node	Node	Likelihood
28	9	5	0.04	82	47	32	0.04	465	35	31	0.01
34	28	5	0.04	138	47	33	0.03	151	50	12	0.02
58	22	5	0.03	140	47	36	0.03	2	51	15	0.18
137	10	5	0.02	88	48	47	0.04	1	51	16	0.2
140	25	5	0.02	11	52	4	0.07	26	51	24	0.1
230	31	28	0.02	61	52	9	0.04	17	51	28	0.12
392	35	34	0.01	4	52	10	0.09	28	51	31	0.1
				269	52	30	0.02	36	51	38	0.07
				18	52	48	0.07	10	51	39	0.14
				19	52	51	0.07	31	51	41	0.09
								7	51	42	0.15
								20	52	51	0.12

According to Table 1 and 2, the regression relationships between aggregation index (u), coefficient of variation (w) (Zhang and Zhan, 2011; Zhang, 2012a), and prediction efficiency ($z=x/y$, where x is the percentages of correctly predicted links, and y is the averaged ranks before which all missing links fall in the list of predicted links), the percentage (%) of correctly predicted links against predicted missing links (q), and the rate of the averaged rank before which all missing links fall in the list of predicted links vs. total number of predicted missing links (f) are as follows

Algorithm prediction:

$$z=0.320+0.344u \quad r^2=0.318, p=0.019<0.05, n=17$$
$$z=0.465+0.192w \quad r^2=0.323, p=0.017<0.05, n=17$$
$$q=1.349+0.243u \quad r^2=0.106, p=0.203, n=17$$
$$q=1.427+0.154u \quad r^2=0.139, p=0.141, n=17$$
$$f=0.438-0.125u \quad r^2=0.306, p=0.021<0.05, n=17$$
$$f=0.389-0.073w \quad r^2=0.341, p=0.014<0.05, n=17$$

Randomization prediction:

$$z=0.485-0.106u \quad r^2=0.149, p=0.125, n=17$$
$$z=0.445-0.063w \quad r^2=0.171, p=0.099<0.1, n=17$$

$$q=1.615-0.349u \quad r^2=0.259, p=0.038<0.05, n=17$$
$$q=1.451-0.182u \quad r^2=0.229, p=0.051<0.01, n=17$$
$$f=0.476-0.088u \quad r^2=0.156, p=0.117, n=17$$
$$f=0.436-0.046w \quad r^2=0.142, p=0.136, n=17$$

Thus prediction efficiency and the percentage of correctly predicted links against predicted missing links with the algorithm increases as the increase of network complexity. Generally, the rate of averaged rank of true missing links in the list of predicted missing links declines as the network complexity, which means the required number for checking true missing links in the predicted list reduces as the increase of network complexity.

Compared to the prediction of randomization method, in general, the results of the algorithm are effective, i.e., the present algorithm is effective in predicting missing links of biological networks (Table 1, 2).

Both mean of node degrees and connectance have not significant relationships with prediction efficiency. Thus prediction efficiency is complexity-depedent only.

Table 2 Link prediction of some tumor related networks of missing links with Rule 1 (*per*=~0.25, λ=1.5).

	PPAR	TGF-β	TNF	STAT3	mTOR	Ras	EGF	PTEN	JAK-STAT
Mean of node degrees	1.85	1.79	2.06	1.75	1.83	1.71	1.96	2.06	2.09
Connectance	0.07	0.05	0.07	0.08	0.04	0.05	0.04	0.06	0.05
Possible maximum number of candidate links	326	669	433	255	993	565	1431	494	858
Aggregation Index (Zhang and Zhan, 2011; Zhang, 2012a)	0.68	0.78	0.85	0.72	0.75	0.75	0.73	0.91	0.91
Coefficient of variation (Zhang and Zhan, 2011; Zhang, 2012a)	0.40	0.61	0.68	0.51	0.54	0.57	0.47	0.82	0.81
Percentage (%) of correctly predicted links against true missing links with the algorithm (x)	**83.3**	**75.0**	**87.5**	**100**	**80.0**	**87.5**	**84.6**	**75.0**	**45.5**
Percentage (%) of correctly predicted links against predicted missing links with the algorithm	**1.9**	**1.3**	**2.0**	**2.6**	**1.4**	**1.8**	**1.3**	**1.7**	**0.9**
Number of missing links	6	8	8	5	10	8	13	8	12
Total number of predicted links with 100 simulations	257	448	346	195	575	392	823	348	545
The averaged rank before which all missing links fall in the list of predicted links (y)	**115**	**190**	**179**	**114**	**202**	**127**	**432**	**47**	**99**
Prediction efficiency (x/y)	**0.7243**	**0.3947**	**0.4888**	**0.8772**	**0.396**	**0.689**	**0.1958**	**1.5957**	**0.4596**
Percentage (%) of correctly predicted links against true missing links with randomization method (x)	**100**	**75**	**87.5**	**60.0**	**70.0**	**37.5**	**61.5**	**100**	**45.5**
Percentage (%) of correctly predicted links against predicted missing links with randomization method	2.2	1.3	1.9	1.4	1.1	0.7	0.9	2.0	0.8
Total number of predicted links with 100 simulations	270	466	375	217	651	424	853	398	617
The averaged rank before which all missing links fall in the list of predicted links (y)	**120**	**148**	**175**	**84**	**338**	**103**	**239**	**302**	**102**
Prediction efficiency (x/y)	**0.8333**	**0.5068**	**0.5**	**0.7143**	**0.2071**	**0.3641**	**0.2573**	**0.3311**	**0.4461**

3.2 Rule 2

In the step (3) of the algorithm, I use the Rule 2 for prediction. The results for some pathways are listed in Table 3. Compared to the Rule 1, the percentages (%) of correctly predicted links with the algorithm calculated from the Rule 2 are overall smaller. However, the prediction efficiency of Rule 2 is generally higher. The major regression relationships and conclusions are similar to Rule 1. Moreover, the prediction efficiency of the algorithm increases dramatically as the network complexity.

Table 2 (continue) Link prediction of some tumor related networks of missing links with Rule 1 ($per=\sim0.25$, $\lambda=1.5$).

	p53	Akt	HGF	JNK	PI3K	MARK	FAS	ERK
Mean of node degrees	1.96	1.69	1.67	2.67	2.25	2.14	1.88	2.27
Connectance	0.04	0.03	0.05	0.06	0.04	0.04	0.04	0.04
Possible maximum number of candidate links	1275	1604	600	1064	1532	1591	1277	1702
Aggregation Index (Zhang and Zhan, 2011; ; Zhang, 2012a)	1.50	3.59	0.96	1.72	0.97	1.22	1.17	1.41
Coefficient of variation (Zhang and Zhan, 2011; Zhang, 2012a)	1.99	5.42	0.93	2.96	0.93	1.46	1.32	1.93
Percentage (%) of correctly predicted links against true missing links with the algorithm (x)	**76.9**	**100**	**57.1**	**100**	**68.8**	**53.3**	**75.0**	**94.1**
Percentage (%) of correctly predicted links against predicted missing links with the algorithm	1.6	2.2	1.1	2.6	1.2	0.9	1.4	1.8
Number of missing links	13	12	7	16	16	14	12	17
Total number of predicted links with 100 simulations	642	542	354	612	899	819	640	883
The averaged rank before which all missing links fall in the list of predicted links (y)	**64**	**66**	**102**	**93**	**240**	**202**	**80**	**179**
Prediction efficiency (x/y)	**1.2016**	**1.5152**	**0.5598**	**1.0753**	**0.2867**	**0.2639**	**0.9375**	**0.5257**
Percentage (%) of correctly predicted links against true missing links with randomization method (x)	**61.5**	**25.0**	**71.4**	**75.0**	**68.8**	**66.7**	**75.0**	**76.5**
Percentage (%) of correctly predicted links against predicted missing links with randomization method	0.9	0.3	1.2	1.4	1.1	1.0	1.2	1.2
Total number of predicted links with 100 simulations	823	862	423	839	990	974	770	1073
The averaged rank before which all missing links fall in the list of predicted links (y)	**343**	**106**	**219**	**296**	**409**	**262**	**177**	**505**
Prediction efficiency (x/y)	**0.1793**	**0.2358**	**0.326**	**0.2534**	**0.1682**	**0.2546**	**0.4237**	**0.1515**

Table 3 Link prediction of some tumor related networks (pathways) of missing links with Rule 2 ($per=\sim0.25$, $\lambda=1.5$).

	Ras	p53	Akt	HGF	JNK	PPAR	TGF-β	TNF
Percentage (%) of correctly predicted links with the algorithm (x)	**62.5**	**92.3**	**50.0**	**57.1**	**75.0**	**33.3**	**37.5**	**62.5**
Percentage (%) of correctly predicted links against predicted missing links with the algorithm	0.7	0.9	0.2	1.5	1.2	1.8	1.5	1.7
Total number of predicted links with 100 simulations	314	388	301	300	404	221	304	291
The averaged rank before which all missing links fall in the list of predicted links (y)	**74**	**92**	**6**	**78**	**81**	**41**	**63**	**95**
Prediction efficiency (x/y)	**0.8446**	**1.0033**	**8.3333**	**0.7321**	**0.9259**	**0.8122**	**0.5952**	**0.6579**
Percentage (%) of correctly predicted links against number of missing links with random network (x)	**37.5**	**53.9**	**16.7**	**85.7**	**62.5**	**83.3**	**87.5**	**75.0**
Percentage (%) of correctly predicted links against predicted missing links with random network	1.6	3.1	2.0	1.3	2.9	0.9	0.9	1.7
Total number of predicted links with 100 simulations	411	823	851	412	838	277	478	359
The averaged rank before which all missing links fall in the list of predicted links (y)	**77**	**325**	**23**	**213**	**246**	**55**	**184**	**111**
Prediction efficiency (x/y)	**0.487**	**0.1658**	**0.7261**	**0.4023**	**0.2541**	**1.5145**	**0.4755**	**0.6757**

4 Discussion

As stated above, random prediction is overall effective for the random networks only. However, in practical applications, most networks are complex networks. Thus the algorithm is effective in predicting missing links in most cases. The prediction efficiency of the algorithm increases as the increase of network complexity. Therefore, the algorithm is more efficient for the networks of higher complexity.

The changes of λ can reflect various effects of the node degree on connection mechanism. The larger λ will lead to find more missing links of the nodes with greater node degree. $\lambda \to 0$ means a trend to random prediction. How to fix a suitable value of λ, is specific to practical problems.

Lü et al. (2015) proposed the structural perturbation method (SPM) to predict missing links and argued that its prediction ability was stronger than previous methods. However, I affim their method does not hold due to the following reasons: (1) Mechanically, the structural perturbation method can only be used to analyze structural stability of dynamic systems. The static structure of a network, expressed by an adjacent matrix, is the topological structure, which cannot represent the dynamic charicteristics of the network evolution. Pediction of missing links should be conducted on the basis of mechanism of network evolution (dynamics). Without loss of generality, network evolution may be approximated with a group of linear differential equations (Zhang, 2015a). And the structural stability of the network was determined by the eigenvalues but not eigenvectors of system matrix. Even so, the structural perturbation method for determining the variables with least impact on structural stability should only be used around the equilibrium states of the system rather than the states far away the equilibrium. (2) During the evolution of a network, the generated links with most likehood are not necessarily those links that minimaly perturb the topological structure of the network. On the premise of not destroying the structural stability of the system and no other limitations, any links will prepare to be created. A most occurred case is that two nodes with most similarity will firstly connect to each other. (3) Utilization of missing links in the prediction model to predict missing links, as done by Lü et al. (2015), is somewhat similar to model fitting but not prediction. In this case, the stronger "prediction" ability (precisely, fitting ability) is surely expected.

So far all prediction methods based on static topological structure only (represented by adjacency matrix) seems to be low efficient. Network evolution based (Zhang, 2012a, 2012c, 2015a, 2016a, 2016b), node similarity based (Zhang, 2015d), and sampling based (correlation based; Zhang, 2007, 2011, 2012b, 2013, 2015b; Zhang and Li, 2015) methods are expected to be the most promising in the future.

Acknowledgment

We are thankful to the support of Discovery and Crucial Node Analysis of Important Biological and Social Networks (2015.6-2020.6), from Yangling Institute of Modern Agricultural Standardization, High-Quality Textbook *Network Biology* Project for Engineering of Teaching Quality and Teaching Reform of Undergraduate Universities of Guangdong Province (2015.6-2018.6), from Department of Education of Guangdong Province, and Project on Undergraduate Teaching Reform (2015.7-2017.7), from Sun Yat-sen University, China.

References

ABCAM. 2012. http://www.abcam.com/index.html?pageconfig=productmap&cl=2282

Amaral LAN. 2008. A truer measure of our ignorance. Proceedings of the National Academy of Sciences of
    USA, 105: 6795-6796

Barabasi AL, Albert R. 1999. Emergence of scaling in random networks. Science, 286(5439): 509

Barzel B, Barabási AL. 2013. Network link prediction by global silencing of indirect correlations. Nature Biotechnology, 31: 720-725

Bastiaens P, Birtwistle MR, Blüthgen N, et al. 2015. Silence on the relevant literature and errors in implementation. Nature Biotechnology, 33: 336-339

Clauset A, Moore C, Newman MEJ. 2008. Hierarchical structure and the prediction of missing links in networks. Nature, 453: 98-101

Guimera R, Sales-Pardo M. 2009. Missing and spurious interactions and the reconstruction of complex networks. Proceedings of the National Academy of Sciences of USA, 106: 22073-22078

Huang JQ, Zhang WJ. 2012. Analysis on degree distribution of tumor signaling networks. Network Biology, 2(3): 95-109

Li JR, Zhang WJ. 2013. Identification of crucial metabolites/reactions in tumor signaling networks. Network Biology, 3(4): 121-132

Lü LY, Medo M, Yeung CH, et al. 2012. Recommender systems. Physics Reports, 519: 1-49

Lü LY, Pan LM, Zhou T, et al. 2015. Toward link predictability of complex networks. Proceedings of the National Academy of Sciences of USA, 112: 2325-2330

Lü LY, Zhou T. 2011. Link prediction in complex networks: A survey. Physica A, 390: 1150-1170

Pathway Central. 2012. SABiosciences. http://www.sabiosciences.com/pathwaycentral.php

Yu HY, Braun P, Yildirim MA, et al. 2008. High-quality binary protein interaction map of the yeast interactome network. Science, 322: 104-110

Zhang WJ. 2007. Computer inference of network of ecological interactions from sampling data. Environmental Monitoring and Assessment, 124: 253-261

Zhang WJ. 2011. Constructing ecological interaction networks by correlation analysis: hints from community sampling. Network Biology, 1(2): 81-98

Zhang WJ. 2012a. Computational Ecology: Graphs, Networks and Agent-based Modeling. World Scientific, Singapore

Zhang WJ. 2012b. How to construct the statistic network? An association network of herbaceous plants constructed from field sampling. Network Biology, 2(2): 57-68

Zhang WJ. 2012c. Modeling community succession and assembly: A novel method for network evolution. Network Biology, 2(2): 69-78

Zhang WJ. 2013. Construction of Statistic Network from Field Sampling. In: Network Biology: Theories, Methods and Applications (WenJun Zhang, ed). 69-80, Nova Science Publishers, New York, USA

Zhang WJ, 2015a. A generalized network evolution model and self-organization theory on community assembly. Selforganizology, 2(3): 55-64

Zhang WJ. 2015b. A hierarchical method for finding interactions: Jointly using linear correlation and rank correlation analysis. Network Biology, 5(4): 137-145

Zhang WJ. 2015c. Calculation and statistic test of partial correlation of general correlation measures. Selforganizology, 2(4): 65-77

Zhang WJ. 2015d. Prediction of missing connections in the network: A node-similarity based algorithm. Selforganizology, 2(4): 91-101

Zhang WJ. 2016a. A random network based, node attraction facilitated network evolution method. Selforganizology, 3(1): 1-9

Zhang WJ. 2016b. Selforganizology: The Science of Self-Organization. World Scientific, Singapore

Zhang WJ, Li X. 2015. Linear correlation analysis in finding interactions: Half of predicted interactions are undeterministic and one-third of candidate direct interactions are missed. Selforganizology, 2(3): 39-45

Zhang WJ, Liu GH. 2012. Creating real network with expected degree distribution: A statistical simulation. Network Biology, 2(3): 110-117

Zhang WJ, Zhan CY. 2011. An algorithm for calculation of degree distribution and detection of network type: with application in food webs. Network Biology, 1(3-4): 159-170

Zhao J, Miao LL, Yang Y, et al. 2015. Prediction of links and weights in networks by reliable routes. Scientific Reports, 5: 12261

Zhou T. 2015. Why link prediction? http://blog.sciencenet.cn/blog-3075-912975.html. Accessed on Aug 14, 2015

13

Unraveling the WRKY transcription factors network in Arabidopsis *Thaliana* by *integrative* approach

Mouna Choura[1], **Ahmed Rebaï**[2], **Khaled Masmoudi**[3]

[1]Laboratory of Plant Protection and Improvement, Center of Biotechnology of Sfax, University of Sfax, Route Sidi Mansour Km 6, P.O.Box 1177, 3018 Sfax, Tunisia

[2]Molecular and Cellular Diagnosis Processes, Center of Biotechnology of Sfax, University of Sfax, Route Sidi Mansour Km 6, P.O.Box 1177, 3018 Sfax, Tunisia

[3]International Center for Biosaline Agriculture (ICBA), P.O.Box 14660, Dubai, United Arab Emirates

E-mail: mouna.choura@cbs.rnrt.tn

Abstract

The WRKY transcription factors superfamily are involved in diverse biological processes in plants including response to biotic and abiotic stresses and plant immunity. Protein-protein interaction network is a useful approach for understanding these complex processes. The availability of *Arabidopsis Thaliana* interactome offers a good opportunity to do get a global view of protein network. In this work, we have constructed the WRKY transcription factor network by combining different sources of evidence and we characterized its topological features using computational tools. We found that WRKY network is a hub-based network involving multifunctional proteins denoted as hubs such as WRKY 70, WRKY40, WRKY 53, WRKY 60, WRKY 33 and WRKY 51. Functional annotation showed seven functional modules particularly involved in biotic stress and defense responses. Furthermore, the gene ontology and pathway enrichment analysis revealed that WRKY proteins are mainly involved in plant-pathogen interaction pathways and their functions are directly related to the stress response and immune system process.

Keywords hub; *in silico*; network; plant immunity; stress; WRKY.

1 Introduction

Plants are frequently exposed to various stresses such as drought, salinity, cold and pathogen attacks. WRKY are large protein family of zinc-finger transcriptional regulators in higher plants involved in biological processes and in biotic and abiotic stress response (Ramamoorthy et al., 2008). They are involved in the regulation of various physiological programs that are unique to plants, including plant pathogen defense, senescence, trichome development (Eulgem, 2000), hormone signaling (Chen et al., 2010; Shang et al., 2010) and secondary metabolism (Wang et al., 2010; Suttipanta et al., 2011). Thus, WRKY proteins play important roles in plant. The WRKY proteins are represented with 74 members in *Arabidopsis* and more than 100 members in rice (Wu et al., 2005; Zhang and Wang, 2005). It is evident that one dimensional annotation is no longer enough to study regulatory proteins that often act with one or more protein partners. Few WRKY – interacting proteins studies were reviewed by Chi et al. (2013) showing the complex regulatory and functional network of WRKY transcription factors and giving insights into biological processes that they do regulate. The availability of the Arabidopsis interactome (Arabidopsis Interactome Mapping Consortium 2011) is encouraging to improve our understanding of WRKY transcription factors family.

Although WRKY proteins were relatively recently discovered class of sequence-specific DNA-binding transcription factors (Rushton et al., 2010), they were extensively studied. Consequently, the literature corpus for WRKY is very large; over 400 articles were retrieved when querying with the generic keyword "WRKY" without counting the large number of protein family members and synonymous.

Our goal is to get a global understanding of comprehensive WRKY protein interactions network in *Arabidopsis thaliana* taking into account available interaction data sources.

For such systems biology analysis, computational approach is needed, where high throughput experiments like yeast two-hybrid (Y2H) analyses, microarrays and text mining of literature are integrated. We then revealed some statistical properties of the network in order to explore the topology and functions of the WRKY network in *Arabidopsis thaliana*.

2 Methods

2.1 Dataset

We queried the Uniprot (release 2014_08) (http://www.uniprot.org/) for WRKY transcription factor in *Arabidopsis thalina* based on (Eulgem, 2000) and we retrieved 74 protein identifiers by excluding hypothetical proteins.

2.2 WRKY-WRKY interaction network construction

The WRKY-WRKY interaction network was performed by version 9.05 STRING (http://string-db.org/) (Franceschini et al. 2013), which is a database of known and predicted protein interactions. The interactions include physical and functional associations derived from genomic context, high-throughput experiments, co-expression and previous knowledge. The interaction data were then assigned by confidence score for each source of evidence.

2.3 Statistical analysis and topological features of WRKY-WRKY network

In a protein-protein interaction (PPI) network, a node denotes a protein and an edge denotes an interaction between two proteins. For each protein set, we applied four topological measures to assess its role in the network: degree, clustering coefficient, betweenness, and shortest-path distance. First, for a node in PPI network, the degree measures the number of links for a node to other nodes. Second, the clustering coefficient of a node is the ratio of the observed number of direct connections between the node's immediate network neighbours over the maximum possible number of such connections. Third, the betweenness of a node is defined as the number of shortest paths between all possible pairs of nodes in the network that traverse the node. Fourth, for a pair of selected nodes in the network, there are many alternative paths between them. The path with the smallest number of links is defined as the shortest path. The number of links passing through in the shortest path is defined as shortest-path distance. We tested whether the interaction network is scale-free or not by plotting the node degree distribution. A network may resemble a scale-free topology if the distribution follows a power-law.

The topological and statistical significance of network have been calculated using Cytoscape plugins Network Analyzer (Assenov et al., 2008) and CentiScaPe (Scardoni et al., 2009).

More explanations of calculated parameters are available at http://med.bioinf.mpi-inf.mpg.de/netanalyzer/ help/2.6.1/index.html.

3 Results and Discussion

3.1 WRKY Network Construction and statistical properties

We integrated results derived from different sources including genomic context, high-throughput experiments, coexpression (conserved), homology among species and text mining related to WRKY transcription factor family proteins of *Arabidopsis thaliana*. We obtained a protein network made up of 72 interactions as described in supplemental material file 1. Among these interactions, only eight are validated experimentally mainly by two-hybrid assay, sixty five protein interactors are coexpressed and all protein pairs are co-cited in bibliography. This finding provides a starting point to validate the other interactions.

By removing disconnected nodes (proteins), advanced view shows highly connected network constituted of 26 WRKY proteins (Fig. 1). Topological parameters of WRKY network are listed in Table 1 (See supplemental material for more details). By plotting the degree distribution in a log scale, we obtained a linear regression R^2 value of 0.35. This is inferring the presence of significant number of negative feedback loops.

Table 1 Network parameters calculated by Network Analyzer.

Number of nodes	26
Number of edges	71
Clustering Coefficient	0.430
Network diamater	6
Network radius	3
Network density	0.218
Network heterogeneity	0.645
Average number of neighbors	5.462
Characteristic path length	2.625

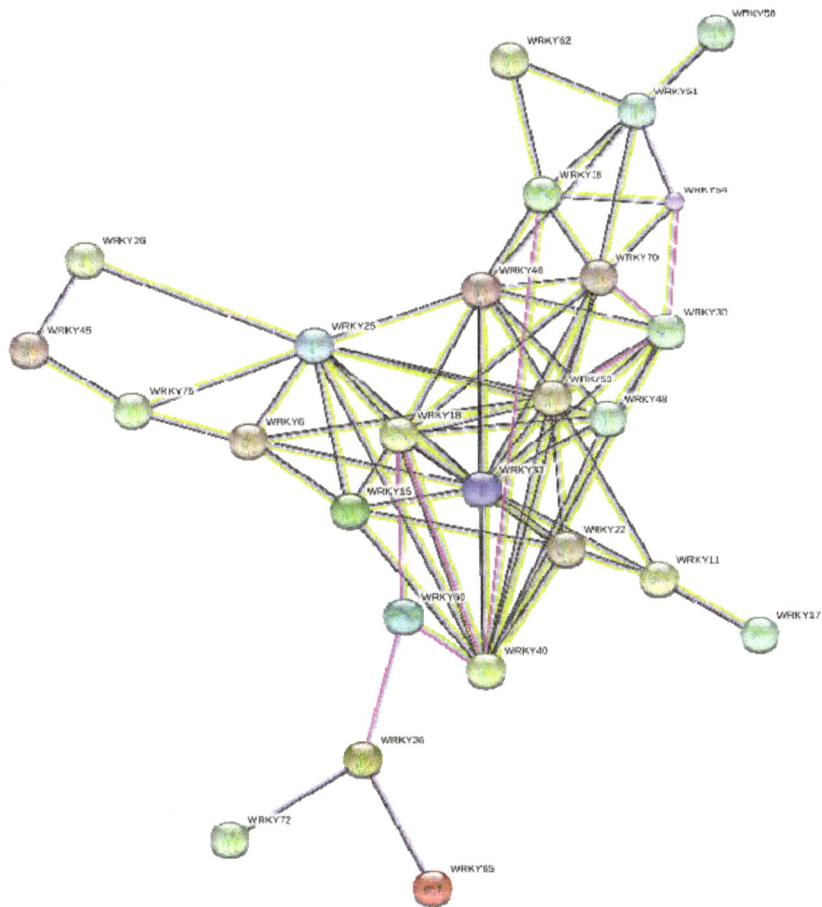

Fig. 1 Evidence view of WRKY proteins network in *Arabidopsis thaliana*.

3.2 Hubs identification

Highly connected nodes are usually defined as 'Hubs'. Based on topological parameters, we have identified protein hubs having often higher degree, clustering coefficient, betweenness and shortest-path distance. By combining these parameters, we have listed the highly ranked nodes in Table 2 (See supplemental material file 2 for complete list). The hubs are thought to maintain network robustness by having higher connectivity in the whole network and may mediate key biological pathways such as signal transduction in the WRKY protein network compared to the other proteins. Our results are in good agreement with previous studies providing strong evidence of WRKY70 role in plant senescence and defense signaling pathways (Ulker et al., 2007) and its cooperation with WRKY54 as negative regulator (Besseau et al., 2012). Moreover, Birkenbihl et al. (2012), revealed the transcriptional regulator role of WRKY33 in *Arabidopsis thaliana* upon *Botrytis cinerea* infection by targeting redox homeostasis, salicylic acid signaling, ethylene jasmonic acid mediated cross communication, and camalexin biosynthesis. Nevertheless, these reports consider that they are still crucial components to identify the full resistance mechanism.

Table 2 Potential hubs in the WRKY network.

Protein	Degree	Clustering coefficient	Betweenness
WRKY40	12	0.5	137
WRKY33	12	0.5	132
WRKY46	10	0.6	125.5
WRKY25	10	0.5	94
WRKY18	10	0.6	75
WRKY70	9	0.6	67.3

3.3 Functional modules in WRKY network

Among the seventy two interactions, we have found seven functional modules (Fig. 2) particularly related to stress response including response to chitin and organic compounds (13 nodes), response to salicylic acid (8 nodes) and defense response to bacterium (13 nodes) and fungus (7 nodes). Additionally, we have found small modules related to calmodulin binding (7 nodes), immune effector process (3 nodes), jasmonic acid mediated pathway (3 nodes) and leaf and organ senescence (3 nodes). Some proteins are present in multiple functional modules such as WRKY 70, WRKY40, WRKY 53, WRKY 60, WRKY 33 and WRKY 51. It could be argued that such proteins are multifunctional and participate in different biological processes. It is worthy to note that these proteins are aforementioned as hubs of the WRKY network. Thus, it would be benefit to further investigate their roles particularly with respect to stress response.

3.4 Pathways and Gene Ontology (GO) enrichment in WRKY network

The enrichment analysis of the network revealed the functional features of WRKY proteins. From the KEGG pathway enrichment, we found that WRKY proteins are included mainly in plant-pathogen interaction pathway (Fisher's exact test, FDR p-value $=1.5 \ 10^{-2}$). (http://www.genome.jp/kegg/pathway/ath/ ath04626.html). Then the enriched GO functions underlying the WRKY proteins are particularly related to response to stress, immune system process, transport and signal transduction (supplementary material file 3). Moreover, this may explain the highly connected architecture of WRKY network. Such system, upon stress, should dynamically respond with the cooperation of all interactors. This is in line with a previous report underlying the role of WRKY transcription factors in plant system immunity (Jones and Dangl, 2006).

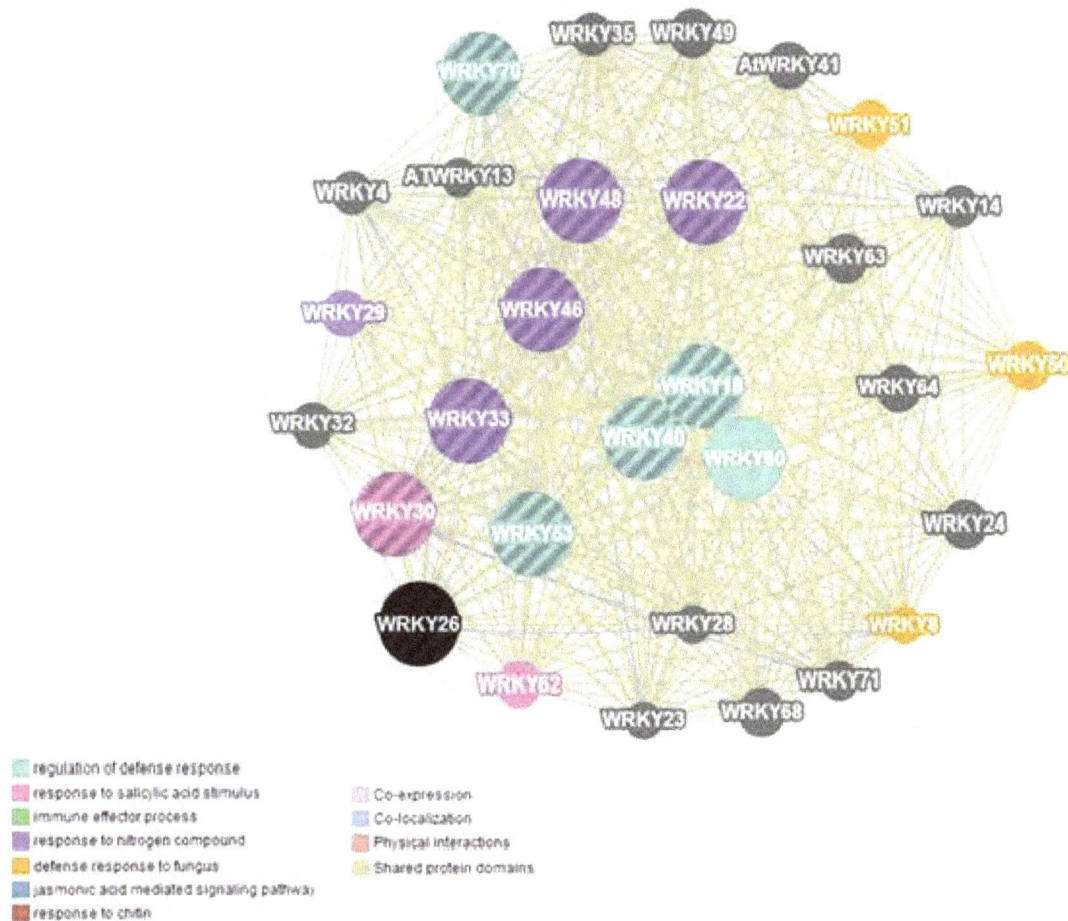

Fig. 2 Functional modules in the WRKY proteins network in *Arabidopsis thaliana.*

4 Conclusions

In this study, we have constructed a WRKY transcription factors network in *Arabidopsis thaliana* using computational tools. By applying integrative approach combining genomic context, high throughput experiments, database content and text mining sources, we get a highly connected network of 72 interactions connected by more than ten hub proteins. This is a fundamental step to uncover plant-pathogen interaction pathway and stress response mediated by WRKY proteins which are not fully established so far. We believe that our approach may be considered as a starting point to address these questions.

Acknowledgements

This work was supported by the Ministry of Higher Education and Scientific Research, Tunisia.

References

Arabidopsis Interactome Mapping Consortium. 2011. Evidence for network evolution in an Arabidopsis interactome map. Science, 333(6042): 601-607

Assenov Y, Ramírez F, Schelhorn SE, et al. 2008. Computing topological parameters of biological networks. Bioinformatics, 24(2): 282-284

Besseau S, Li J, Palva ET. 2012. WRKY54 and WRKY70 co-operate as negative regulators of leaf senescence in Arabidopsis thaliana. Journal of Experimental Botany, 63(7): 2667-2679

Birkenbihl RP, Diezel C, Somssich IE. 2012. Arabidopsis WRKY33 is a key transcriptional regulator of hormonal and metabolic responses toward Botrytis cinerea infection. Plant Physiology, 159(1): 266-285

Chen H, Lai Z, Shi J, Xiao Y, Chen Z, Xu X. 2010. Roles of Arabidopsis WRKY18, WRKY40 and WRKY60 transcription factors in plant responses to abscisic acid and abiotic stress. BMC Plant Biology, 10: 281

Chi Y, Yang Y, Zhou Y, et al. 2013. Protein-Protein interactions in the regulation of WRKY transcription factors. Molecular Plant, 6(2): 287-300

Eulgem T, Rushton PJ, Robatzek S, Somssich IE. 2000. The WRKY superfamily of plant transcription factors. Trends in Plant Science, 5(5): 199-206

Franceschini A, Szklarczyk D, Frankild S, et al. 2013. STRING v9.1: protein-protein interaction networks, with increased coverage and integration. Nucleic Acids Research, 41(D1): D808-D815

Jones J D G, Dangl J L. 2006. The plant immune system. Nature, 444(7117): 323-329

Ramamoorthy R, Jiang SY, Kumar N, Venkatesh PN, Ramachandran S. 2008. A comprehensive transcriptional profiling of the WRKY gene family in rice under various abiotic and phytohormone treatments. Plant and Cell Physiology, 49(6): 865-879

Rushton PJ, Somssich IE, Ringler P, Shen QJ. 2010. WRKY transcription factors. Trends in Plant Science, 15 (5): 247-258

Scardoni G, Petterlini M, Laudanna C. 2009. Analyzing biological network parameters with CentiScaPe. Bioinformatics, 25(21): 2857-2859

Shang Y, Yan L, Liu ZQ, et al. 2010. The Mg-chelatase H subunit of Arabidopsis antagonizes a group of WRKY transcription repressors to relieve ABA-responsive genes of inhibition. Plant Cell, 22(6): 1909-1935

Suttipanta N, Pattanaik S, Kulshrestha M, et al. 2011. The transcription factor CrWRKY1 positively regulates the terpenoid indole alkaloid biosynthesis in Catharanthus roseus. Plant Physiology, 157(4): 2081-2093

Ulker B, Shahid Mukhtar M, Somssich IE. 2007. The WRKY70 transcription factor of Arabidopsis influences both the plant senescence and defense signaling pathways. Planta, 226(1): 125-137

UniProt Consortium. 2013. Update on activities at the Universal Protein Resource (UniProt) in 2013. 2014. Nucleic Acids Research, 41(D1): D43-D47

Wang H, Avci U, Nakashima J, et al. 2010. Mutation of WRKY transcription factors initiates pith secondary wall formation and increases stem biomass in dicotyledonous plants. Proceedings of the National Academy of Sciences of USA, 107(51): 22338–22343

Wu KL, Guo ZJ, Wang HH, Li J. 2005. The WRKY family of transcription factors in rice and Arabidopsis and their origins. DNA Research, 12(1): 9-26

Zhang Y, Wang L. 2005. The WRKY transcription factor superfamily: its origin in eukaryotes and expansion in plants. BMC Evolutionary Biology, 5: 1

14

Regression analysis on different mitogenic pathways

Shruti Jain

Department of Electronics and Communication Engineering, Jaypee University of Information Technology, Solan-173234, India

E-mail: jain.shruti15@gmail.com

Abstract

In this paper different regression analysis methods were discussed on three different mitogenic pathways i.e. ERK, MK2 and JNK. Coefficient of determination, ANOVA, T-value, Durban-Watson statistics were calculated for the corresponding three proteins. The model was made using linear modeling using different regression analysis techniques in which different parameters like Mean sq error, Root mean sq error, Mean abs error, Relative sq error, Root relative sq error and Relative abs error were calculated using different analysis like PLS, linear, SVM, random forest etc were calculated. In all respect results with ERK are the best.

Keywords regression analysis; ERK; JNK; MK2.

1 Introduction

MAP kinases are actually a family of protein kinases (Jain, 2012; Janes et al., 2005; Suzanne et al., 2005; Weiss, 2001) that are widely distributed and are found in all eukaryotic organisms. The MAPKs is divided (Jain et al., 2009; Kyriakis et al., 1996; Pearson et al., 2001) into three families as Extracellular-regulated kinase (ERK) (Jain et al., 2010), p38/high osmolarity glycerol (HOG)/Mitogen-activated protein kinase 2 (MK2) (Jain et al., 2010), stress-activated protein kinase (JNK/SAPK) (Jain, et al 2010; Jain, 2014), which leads to cell death/ cell survival. The ERK are activated by differentiation and mitogenic signals while JNK & MK2 are respond to stress and inflammaton.

ERK pathway is activated by the binding of SOS with Grb2 which results in SOS. SOS then activates RAS which leads RAF. RAF activates MEK1 and MEK2 which further activates ERK1 and ERK 2 respectively (Jain et al., 2010). The second most widely studied MAP kinase cascade is the c-Jun NH2-terminal kinase/stress activated protein kinase (JNK/ SAPK) (Brockhaus et al., 1990; Jain, 2014). The JNK are activated when cells are exposed to ultraviolet radiation, heat shock, or inflammatory cytokines. The p38/ MK2 kinase is the well-characterized member of the MAP kinase family. It is activated in response to inflammatory cytokines, endotoxins, and osmotic stress.

In this paper we will study regarding the regression analysis of different MAPK pathways. We have calculated the mean sq error, root mean sq error, mean abs error, relative sq error, root relative sq error and relative abs error using different analysis like PLS, linear, SVM, random forest etc. Coefficient of determination, ANOVA, T-value, Durban-Watson statistics were also calculated for the corresponding three proteins

2 Materials and Methods

In this paper we are using different regression analysis methods on different MAPK pathways i.e., ERK, MK2 and JNK. Ten different concentrations for TNF (Brockhaus et al., 1990; Jain et al., 2011), EGF (Jain, 2015; Libermann et al., 1984; Normanno et al., 2006) and Insulin (Jain, 2011, 2015; Lizcano et al., 2002) were used in ng/ml, i.e., 0-0-0, 5-0-0, 100-0-0, 0-100-0, 5-1-0, 100-100-0, 0-0-500, 0.2-0-1, 5-0-5, 100-0-500.

A) Calculation of coefficient of determination and Durban Watson statistics:

We have calculated the values of coefficient of determination and Durban Watson statistics for three proteins (ERK, MK2 and JNK) which are the main proteins for MAPK pathway. Equation 1 gives the regression coefficient (r^2) equation.

$$r^2 = 1 - \frac{\sum(y_i - f_i)^2}{\sum(y_i - \overline{y})^2} \quad or \quad r^2 = \frac{\sum(f_i - \overline{f})^2}{\sum(y_i - \overline{y})^2}$$

(1)

where y_i are the observed values, f_i are the predicted values, $\overline{y}$ are the mean values of observed data and $\overline{f}$ are the mean values of the predicted data. If the regression model is perfect, error sum of square (SSE) is equal to zero, and its r^2 is 1. If the regression model is not perfect, SS_E is equal to SS_T and its r^2 is zero. Inputs are ten different concentrations of TNF- EGF - Insulin (0-0-0, 5-0-0, 100-0-0, 0-100-0, 5-1-0, 100-100-0, 0-0-500, 0.2-0-1, 5-0-5, 100-0-500) All are ng/ml. Outputs are ERK or JNK or MK2 values.

For our data sets of ten concentrations of TNF/EGF/ Insulin for ERK pathway values are as: S = 0.005931, regression coefficient (r^2) = 92.2%, $r^2_{(adj)}$ = 92.0%, $r^2_{(pred)}$ = 91.6%, PRESS = 0.010980, Durbin-Watson Statistics =2.04.

For our data sets of ten concentrations of TNF/EGF/ Insulin for MK2 pathway values are as: S = 0.006480, r^2 = 90.7% , $r^2_{(adj)}$ = 90.4%, $r^2_{(pred)}$ = 90.01%, PRESS = 0.013055, Durbin Watson Statistics = 2.09.

For our data sets of ten concentrations of TNF/EGF/ Insulin for JNK are as: S = 0.01042, (r^2) = 76.0%, adjusted regression coefficient ($r^2_{(adj)}$) = 75.2%, $r^2_{(pred)}$ = 74.19%, PRESS = 0.033734, Durbin Watson Statistics = 1.61.

B) Calculation of Analysis of Variance (ANOVA) (Table 1, 2 and 3):

Table 1 Analysis of Variance (ANOVA) for all combinations of ERK.

Source	df	SS	MS	F
Regression	10	0.120516 (SSa)	0.012052 (MSa)	342.60
Residual Error	289	0.010166 (SSe)	0.000035 (MSe)	
Total	299	0.130683		

Table 2 Analysis of Variance (ANOVA) for all combinations of MK2.

Source	df	SS	MS	F
Regression	10	0.118548 (SSa)	0.011855 (MSa)	282.33
Residual Error	289	0.012135 (SSe)	0.000042 (MSe)	
Total	299	0.130683		

Table 3 Analysis of Variance (ANOVA) for all combinations of JNK.

Source	df	SS	MS	F
Regression	10	0.099332 (SSa)	0.009933 (MSa)	91.57
Residual Error	289	0.031350 (SSe)	0.000108 (MSe)	
Total	299	0.130682		

Table 1, Table 2 and Table 3 shows the mean squares of the regression, sum of squares and residual error for ERK, MK2 & JNK respectively. Table 4, Table 5 & Table 6 shows the standard error coefficient value, T-value, p- value, f-value which was also calculated for ERK, MK2 & JNK respectively. In these tables, SS is the sum of square, df is the degree of freedom, MS is the mean square, SS_a is the sum of square among groups, SS_e is the error sum of square, MS_a is the average variability among groups, MS_e is the average variability within groups. They are calculated as: $MS_a = SS_a/\text{df}$, $MS_e = SS_e/\text{df}$, $F = MS_a/MS_e$

Table 4 Regression analysis in terms of standard error coefficients, p value, T-value, F- value for ERK.

Effect	Coefficient	Standard Error Coefficient	t-value	p- value	VIF (variance inflation factor)
Constant	0.34684	0.02003	17.32	0.000	
0-0-0	0.00013307	0.00005417	2.46	0.015	35.0
5-0-0	0.00014586	0.00007029	2.08	0.039	8.7
100-0-0	0.00020099	0.00007236	2.78	0.006	97.1
0-100-0	0.00009306	0.00005560	1.67	0.095	21.4
5-1-0	0.00001065	0.00006060	0.18	0.861	10.6
100-100-0	0.00001734	0.00006432	0.27	0.788	77.3
0-0-500	0.00010723	0.00007165	1.50	0.136	77.4
0.2-0-1	0.00009634	0. 00004255	2.26	0.024	129.5
5-0-5	0.00004758	0.00001973	2.41	0.016	247.3
100-0-500	0.00001964	0.00004623	0.42	0.671	168.9

Table 5 Regression analysis in terms of standard error coefficients, p value, T-value, F- value for MK2.

Effect	Coefficient	Standard Error Coefficient	t-value	p- value	VIF (variance inflation factor)
Constant	0.59869	0.03739	16.01	0.000	
0-0-0	0.00002453	0.00005234	0.47	0.640	1.7
5-0-0	-0.00004073	0.00005098	-0.80	0.425	2.1
100-0-0	-0.00033461	0.00003862	-8.67	0.000	13.7
0-100-0	0.00000636	0.00004956	0.13	0.898	2.0
5-1-0	0.00010052	0.00005540	1.81	0.071	4.1
100-100-0	-0.00001375	0.00004750	-0.29	0.772	23.2
0-0-500	0.00003543	0.00005259	0.67	0.501	15.5
0.2-0-1	0.00009931	0.00004462	2.23	0.027	9.9
5-0-5	-0.00027582	0.00004681	-5.89	0.000	18.9
100-0-500	-0.00007383	0.00004397	-1.68	0.094	33.0

Table 6 Regression analysis in terms of standard error coefficients, p value, T-value, F- value for JNK.

Effect	Coefficient	Standard Error Coeff	T-Value	p- Value	F- Value
0-0-0	0.92309	0.04895	18.86	0.000	
5-0-0	0.00012979	0.00008235	1.58	0.116	1.6
100-0-0	-0.00023739	0.00008791	-2.70	0.007	1.5
0-100-0	-0.00075872	0.00006467	-11.73	0.000	1.6
5-1-0	-0.00000557	0.00008274	-0.07	0.946	1.0
100-100-0	-0.00038617	0.00007818	-4.94	0.000	1.8
0-0-500	0.00009802	0.00008049	1.22	0.224	1.2
0.2-0-1	0.00004452	0.00008060	0.55	0.581	1.0
5-0-5	-0.00042354	0.00008249	-5.13	0.000	4.2
100-0-500	-0.00004693	0.00008336	-0.56	0.574	1.6

We clubbed all the concentrations of TNF, EGF and Insulin and only normalized output (ERK, MK2, JNK) were taken and we get the regression equation as:

Final Output for ERK = 0.347 +0.000133 a +0.000146 b +0.000201 c +0.000093 d +0.000011 e +0.000017 f +0.000107 g +0.000096 h +0.000048 i +0.000020 j   (2)

Final Output for MK2 = 0.599 +0.000025 a-0.000041 b -0.000335 c +0.000006 d +0.000101 e -0.000014 f +0.000035 g +0.000099 h -0.000276 i -0.000074 j   (3)

Final Output for JNK =0.923 + 0.000130a - 0.000237b- 0.000759c - 0.000006d- 0.000386e + 0.000098f +0.000045g-0.000424h-0.000047i -0.000154j.   (4)

where a, b, c are the different concentrations of TNF, EGF and Insulin.

We have calculated mean sq error (MSE), root mean sq error (RMSE), mean abs error (MAE), relative sq error (RSE), root relative sq error (RRSE) and relative abs error (RAE) for ERK, MK2 and JNK using different regression analysis like PLS, linear, SVM, KNN, random forest, regression etc was given in Table 7, Table 8 and Table 9 respectively.

- Mean square error: where fi is the predicted value and yi is the actual/ observed value

$$MSE = \frac{1}{n} \sum_{i=1}^{n} (y_i - f_i)^2 \tag{5}$$

- Root mean sq error: RMSE is used to measure the error rate of a regression model. However, it can only be compared between models whose errors are measured in the same units.

$$RMSE = \sqrt{\frac{1}{n} \sum_{i=1}^{n} (y_i - f_i)^2} \tag{6}$$

- Mean abs error: MAE has the same unit as the original data, and it can only be compared between models whose errors are measured in the same units. It is usually similar in magnitude to RMSE, but slightly smaller.

$$MAE = \frac{1}{n} \sum_{i=1}^{n} |y_i - f_i| \tag{7}$$

- Relative sq error: RSE can be compared between models whose errors are measured in the different units.

$$RSE = \frac{\sum_{i=1}^{n} (y_i - f_i)^2}{\sum_{i=1}^{n} (\bar{y} - f_i)^2} \tag{8}$$

- Root relative sq error:

$$RRSE = \sqrt{\frac{\sum_{i=1}^{n} (y_i - f_i)^2}{\sum_{i=1}^{n} (\bar{y} - f_i)^2}} \tag{9}$$

- Relative abs error: RAE can be compared between models whose errors are measured in the different units.

$$RAE = \frac{\sum_{i=1}^{n} |y_i - f_i|}{\sum_{i=1}^{n} |\bar{y} - f_i|} \tag{10}$$

Table 7 Various parameters using different regression methods for ERK.

	MSE	RMSE	MAE	RSE	RRSE	RAE
PLS Regression	0.0000	0.0060	0.0048	0.0836	0.2891	0.2551
Linear Regression	0.0000	0.0060	0.0048	0.0836	0.2891	0.2551
SVM Regression	0.0005	0.0216	0.0207	1.0755	1.0371	1.0969
K nearest neighbours regression	0.0001	0.0072	0.0058	0.1189	0.3448	0.3089
Mean	0.0004	0.0210	0.0190	1.0125	1.0062	1.0042
Random Forest regression	0.0000	0.0065	0.0052	0.0964	0.3105	0.2727
Regression tree	0.0000	0.0060	0.0048	0.0840	0.2898	0.2557

Table 8 Various analysis parameters using diff regression methods for MK2.

	MSE	RMSE	MAE	RSE	RRSE	RAE
PLS Regression	0.0000	0.0067	0.0054	0.1017	0.3190	0.2847
Linear Regression	0.0000	0.0067	0.0054	0.1017	0.3190	0.2847
SVM Regression	0.0005	0.0216	0.0207	1.0755	1.0371	1.0969
K nearest neighbours regression	0.0001	0.0072	0.0058	0.1187	0.3445	0.3058
Mean	0.0004	0.0210	0.0190	1.0125	1.0062	1.0042
Random Forest regression	0.0000	0.0070	0.0055	0.1113	0.3336	0.2888
Regression tree	0.0000	0.0060	0.0048	0.0840	0.2898	0.2557

Table 9 Various analysis parameters using diff regression methods for JNK.

	MSE	RMSE	MAE	RSE	RRSE	RAE
PLS Regression	0.0001	0.0106	0.0085	0.2599	0.5098	0.4521
Linear Regression	0.0001	0.0106	0.0085	0.2599	0.5098	0.4521
SVM Regression	0.0005	0.0216	0.0207	1.0755	1.0371	1.0969
K nearest neighbors regression	0.0001	0.0093	0.0066	0.2004	0.4477	0.3517
Mean	0.0004	0.0210	0.0190	1.0125	1.0062	1.0042
Random Forest regression	0.0001	0.0088	0.0067	0.1763	0.4199	0.3552
Regression tree	0.0000	0.0069	0.0051	0.1104	0.3323	0.2692

3 Conclusion

In this paper we have discussed the coefficient of determination, ANOVA, T-value, Durban Watson statistics, P-value, F-value, standard error coefficients for different mitogenic pathways, i.e., JNK, ERK, and MK2. We have also calculated MSE, RMSE, MAE, RSE, etc., using different analysis methods like PLS, linear, SVM, random forest etc. In all respect results with ERK are the best. In future we will find the pdf all proteins using different distribution functions.

References

Brockhaus M, Schoenfeld HJ, Schlaeger EJ, Hunziker W, Lesslauer W, Loetscher H. 1990. Identification of two types of tumor necrosis factor receptors on human cell lines by monoclonal antibodies. Proceedings of the National Academy of Sciences USA, 87: 3127-3131

Jain S, Naik PK, Sharma R. 2009. A computational modeling of cell survival/ death using VHDL and MATLAB simulator. Digest Journal of Nanomaterials and Biostructures (DJNB), 4(4): 863- 879

Jain S, Bhooshan SV, Naik PK. 2010. Model of mitogen activated protein kinases for cell survival/death and its equivalent bio-circuit. Current Research Journal of Biological Sciences, 2(1): 59-71

Jain S, Bhooshan SV, Naik PK. 2011, Mathematical modeling deciphering balance between cell survival and cell death using insulin. Network Biology, 1(1): 46-58

Jain S, Bhooshan SV, Naik PK. 2011. Mathematical modeling deciphering balance between cell survival and cell death using tumor necrosis factor α. Research Journal of Pharmaceutical, Biological and Chemical Sciences, 2(3): 574-583

Jain S. 2012. Communication of signals and responses leading to cell survival / cell death using Engineered Regulatory Networks. PhD Thesis, Jaypee University of Information Technology, Solan, Himachal Pradesh, India

Jain S. 2014. Implementation of fuzzy system using operational transconductance amplifier for ERK pathway of EGF/ Insulin leading to cell survival/ death. Journal of Pharmaceutical and Biomedical Sciences, 4(8): 701-707

Jain S, Chauhan DS. 2015. Mathematical analysis of receptors for survival proteins. International Journal of Pharmaceutical and Biomedical Sciences, 6(3): 164-176

Jain S, Chauhan DS. 2015. Implementation of fuzzy system using different voltages of OTA for JNK pathway leading to cell survival/ death. Network Biology, 5(2): 62-70

Janes KA, John AG, Suzanne G, Peter SK, Douglas LA, Michael YB. 2005. A systems model of signaling identifies a molecular basis set for cytokine-induced apoptosis; Science, 310: 1646-1653

Kyriakis JM, Avruch J. 1996. Sounding the alarm: protein kinase cascades activated by stress and inflammation. Journal of Biological Chemistry, 271: 24313-24316

Libermann TA , Razon TA., Bartal AD, Yarden Y., Schlessinger J, Soreq H. 1984. Expression of epidermal growth factor receptors in human brain tumors Cancer Research 44:753-760

Lizcano JM, Alessi DR. 2002. The insulin signalling pathway. Current Biology, 12: 236-238

Normanno N, De Luca A, Bianco C, Strizzi L, Mancino M, Maiello MR,, Carotenuto A, De Feo G, Caponiqro F, Salomon DS. 2006. Epidermal growth factor receptor (EGFR) signaling in cancer. Gene, 366: 2-16

Pearson G, Robinson F, Beers GT, Xu BE, Karandikar M, Berman K, Cobb MH. 2001. Mitogen-activated protein (MAP) kinase pathways: regulation and physiological functions. Endocrine Reviews, 22(2): 153-183

Suzanne G, Janes KA, John AG, Emily PA, Douglas LA, Peter SK. 2005. A compendium of signals and responses trigerred by prodeath and prosurvival cytokines. Manuscript M500158-MCP200

Weiss R. 2001. Cellular computation and communications using engineered genetic regulatory networks. PhD Thesis, MIT, USA

15

Stability analysis of a biological network

Q. Din
Department of Mathematics, University of Poonch Rawalakot, Pakistan
Email: qamar.sms@gmail.com

Abstract
In this paper, we study qualitative behavior of a network of two genes repressing each other. More precisely, we investigate the boundedness character and persistence, existence and uniqueness of positive steady-state, local asymptotic stability and global behavior of unique positive equilibrium point of this model.

Keywords biological network; stability analysis; local asymptotic stability; global behavior.

1 Introduction

In case of systems biology it is very crucial impression of modeling the qualitative behavior of biological and biochemical networks where molecules are represented as nodes and the molecular interactions are so called edges. Due to scope and complicated behavior of these networks it is very important to discuss and study their dynamical behavior. An interaction dynamics can be used instead of an explicit mathematical description of these biological networks and computer simulations can be used to study the dynamical behavior of these complex biological networks. It is well known fact that dynamics is related to the study of changes with respect to time. For example in case of classical mechanics an apple falling to the ground, or the growth of the human population. Particularly, in case of systems biology dynamics is related to the changes in concentrations of molecules (or numbers) within a cell. Differential equations and difference equations are main tools for modeling these biological networks.

A dynamical system is defined by a set of variables describing the state of the system and the laws for which the values of these variables change with respect to time. Variables can be regarded as discrete-time variables where the state of the variable can be described by a distinct set of values, or continuous variables in which any real value can be used. The option of differential equations or difference equations depends upon the time and on the state of all variables. Furthermore, it can be deterministic where the time and variable states uniquely defines the state at next time point, or it can be stochastic where the time and variable state defines the probability of how the variable values changes over time. The goal when dealing with a dynamical system is to describe and analyze the behavior of the individual variables and also of the complete system, and to be able to make predictions. A dynamical system can be in equilibrium where variables do not change, it can oscillate in a repeating fashion, or it can be more complicated and even chaotic.

In this paper, we study the qualitative behavior of construction of a genetic toggles witch in *Escherichia coli*. The dynamical behavior of the toggle switch can be described by using the following planar system of nonlinear differential equations:

$$\frac{dx}{dt} = \frac{a}{1+y^\alpha} - x, \quad \frac{dy}{dt} = \frac{b}{1+x^\beta} - y, \tag{1.1}$$

where x, y are concentrations of the two repressors, a, b are the rates of synthesis of repressors, respectively. Moreover, α, β are cooperativity factors. For further detail of system (1.1) we refer interested readers to (Gardner et al., 2000). As it is pointed out in (Zhou and Zou, 2003; Liu, 2010) the discrete time models governed by difference equations are more appropriate than the continuous ones when the populations are of non-overlapping generations. The study of discrete-time models described by difference equations has now been paid great attention since these models are more reasonable than the continuous time models when populations have non-overlapping generations. Discrete-time models give rise to more efficient computational models for numerical simulations and also show rich dynamics compared to the continuous ones (Ahmad, 1993; Tang and Zou, 2006). It is very interesting mathematical problem to discuss qualitative behavior of discrete dynamical systems. For more results for the qualitative behavior of discrete dynamical systems, we refer the reader to (Papaschinopoulos et al., 2011; Din, 2013, 2014; Din and Donchev, 2013).

Using the Euler's method the discretization of (1.1) can be obtained, where the discretization preserves the property of convergence to the equilibrium, regardless of the step size.

Let $t_n = nh$, where h is step size. Applying Euler's method, we obtain

$$x_{n+1} = \frac{ah}{1+(y_n)^\alpha} + (1-h)x_n, \quad y_{n+1} = \frac{bh}{1+(x_n)^\beta} + (1-h)y_n. \tag{1.2}$$

Re-scaling the parameters in (1.2) by $ah \to A, bh \to B, 1-h \to C$, we obtain the following discrete dynamical system:

$$x_{n+1} = \frac{A}{1+(y_n)^\alpha} + Cx_n, \quad y_{n+1} = \frac{B}{1+(x_n)^\beta} + Cy_n. \tag{1.3}$$

In this paper, our aim is to study the boundedness and persistence, existence and uniqueness of positive equilibrium point, local asymptotic stability and global asymptotic behavior of unique positive equilibrium point of discrete dynamical system (1.3).

2 Main Results

Theorem 2.1: Assume that $C < 1$, then every positive solution of (1.3) is bounded and persists.

Proof: Let $\{(x_n, y_n)\}$ be any arbitrary positive solution of (1.3), then one has

$$x_{n+1} \le A + Cx_n, \quad y_{n+1} \le B + Cy_n$$

for all $n = 0,1,2,\cdots$. Furthermore, consider the following system of linear difference equations:

$$u_{n+1} = A + Cu_n, \quad v_{n+1} = B + Cv_n$$

for all $n = 0,1,2,\cdots$. Then solution of this linear system is given by

$$u_n = \frac{A(1-C^n)}{1-C} + C^n u_0, v_n = \frac{B(1-C^n)}{1-C} + C^n v_0$$

for all $n = 1,2,\cdots$, and where u_0, v_0 are initial conditions. Assume that $C < 1$, then it follows that

$$u_n \le \frac{A}{1-C} + u_0, v_n \le \frac{B}{1-C} + v_0$$

for all $n = 1,2,\cdots$. Taking $u_0 = x_0, v_0 = y_0$, then it follows by comparison that

$$x_n \leq \frac{A}{1-C}, y_n \leq \frac{B}{1-C}$$

for all $n = 1,2,\cdots$. Furthermore, it follows from (1.3) that

$$x_{n+1} \geq \frac{A}{1+(y_n)^\alpha} \geq \frac{A}{1+\left(\frac{B}{1-C}\right)^\alpha} = K_1,$$

$$y_{n+1} \geq \frac{B}{1+(x_n)^\alpha} \geq \frac{B}{1+\left(\frac{A}{1-C}\right)^\alpha} = K_2.$$

Hence, we obtain

$$K_1 \leq x_n \leq \frac{A}{1-C}, K_2 \leq y_n \leq \frac{B}{1-C}$$

for all $n = 1,2,\cdots$. This completes the proof.

.

Theorem 2.2: Assume that $C < 1$, then for every positive solution of (1.3) the set $\left[K_1,\frac{A}{1-C}\right] \times \left[K_2,\frac{B}{1-C}\right]$ is an invariant set.

Proof: Let $\{(x_n, y_n)\}$ be any arbitrary positive solution of (1.3) such that the initial conditions $(x_0, y_0) \in \left[K_1,\frac{A}{1-C}\right] \times \left[K_2,\frac{B}{1-C}\right]$ Then it follows from system (1.3) that

$$K_1 \leq x_1 = \frac{A}{1+(y_0)^\alpha} + Cx_0 \leq \frac{A}{1-C}, \quad K_2 \leq y_1 = \frac{B}{1+(x_0)^\beta} + Cy_0 \leq \frac{B}{1-C}.$$

From mathematical induction, we obtain that

$$(x_n, y_n) \in \left[K_1,\frac{A}{1-C}\right] \times \left[K_2,\frac{B}{1-C}\right]$$

for all $n = 1,2,\cdots$.

Theorem 2.3: Suppose that $C < 1$, then (1.3) has unique positive equilibrium point $(\bar{x}, \bar{y}) \in \left[K_1,\frac{A}{1-C}\right] \times \left[K_2,\frac{B}{1-C}\right]$, if the following conditions are satisfied:

$$K_1 < \frac{B}{(1-C)\left[1+(f(K_1))^\alpha\right]},$$

$$\alpha\beta\left(\frac{A}{1-C}\right)^\beta [A - (1-C)K_1] < A\left(1 + K_1^\beta\right),$$

where $K_1 = \frac{A}{1+\left(\frac{B}{1-C}\right)^\alpha}$ and $f(x) = \frac{B}{(1-C)(1+x^\beta)}$.

Proof: Consider the following system of algebraic equations

$$x = \frac{A}{1+(y)^\alpha} + Cx, \quad y = \frac{B}{1+(x)^\beta} + Cy. \tag{2.1}$$

Then, it follows from (2.1) that $x = \frac{A}{(1-C)(1+y^\alpha)}$ and $y = \frac{B}{(1-C)(1+x^\beta)}$. Assume that

$$(x, y) \in \left[K_1,\frac{A}{1-C}\right] \times \left[K_2,\frac{B}{1-C}\right].$$

Set

$$F(x) = \frac{A}{(1-C)\left[1+(f(x))^\alpha\right]} - x,$$

where $f(x) = \frac{B}{(1-C)(1+x^\beta)}$. Then it follows that $F(K_1) = \frac{B}{(1-C)\left[1+(f(K_1))^\alpha\right]} - K_1 > 0$ for all $C < 1$.

Furthermore, $F\left(\frac{A}{1-C}\right) = \dfrac{A}{(1-C)\left[1+\left(\dfrac{B}{\left(1+\left(\frac{A}{1-C}\right)^\beta\right)(1-c)}\right)^\alpha\right]} - \dfrac{A}{1-C} < 0$ if and only if

$\dfrac{1}{1+\left(\dfrac{B}{\left(1+\left(\frac{A}{1-c}\right)^\beta\right)(1-c)}\right)^\alpha} < 1$, that is, $\left(\dfrac{B}{\left(1+\left(\frac{A}{1-c}\right)^\beta\right)(1-c)}\right)^\alpha > 0$. Hence, $F\left(\frac{A}{1-C}\right) < 0$ for all $C < 1$.

Thus $F(x)$ has a root in $\left[K_1, \frac{A}{1-C}\right]$. Furthermore, we have

$$F'(x) = -1 - \frac{A\alpha f(x)^{-1+\alpha} f'(x)}{(1-C)(1+f(x)^\alpha)^2},$$

$$f'(x) = -\frac{Bx^{-1+\beta}\beta}{(1-C)(1+x^\beta)^2}.$$

Let z be a solution of $F(x) = 0$, then $z = \frac{A}{(1-C)\left[1+(f(z))^\alpha\right]}$. Hence it follows that

$$F'(z) = -1 + \frac{Az^{-1+\beta}\left(\dfrac{B}{(1-C)(1+z^\beta)}\right)^{1+\alpha}\alpha\beta}{B\left(1+\left(\dfrac{B}{(1-C)(1+z^\beta)}\right)^\alpha\right)^2}$$

$$= -1 + \frac{\alpha\beta z^\beta(A-(1-C)z)}{A(1+z^\beta)}$$

$$< -1 + \frac{\alpha\beta\left(\dfrac{A}{1-C}\right)^\beta[A-(1-C)K_1]}{A\left(1+K_1^\beta\right)} < 0,$$

which completes the proof.

Lemma 2.1 (Sedaghat, 2003): Assume that $X_{n+1} = F(X_n)$, $n = 0,1,...$ be a system of difference equations and $\bar{X}$ is an equilibrium point of F. If all eigenvalues of the Jacobian matrix J_F about the fixed point $\bar{X}$ lie inside the open unit disk $|\lambda| < 1$, then $\bar{X}$ is locally asymptotically stable. If one of them has absolute value greater than one, then $\bar{X}$ is unstable.

Lemma 2.2 (Grove and Ladas, 2004): Consider the following equation

$\lambda^2 + a\lambda + b = 0,$ (2.2)

where a and b are real numbers. Then, the necessary and sufficient condition for both roots of the equation (2.2) to lie inside the open disk $|\lambda| < 1$ is

$|a| < 1 + b < 2.$

Theorem 2.4: The unique positive equilibrium point of system (1.3) is locally asymptotically stable, if the following condition is satisfied:

$$2C + C^2 + \frac{\alpha\beta A^\beta B^\alpha}{(1-C)^{\alpha+\beta-2}} < 1.$$

Proof. The Jacobian matrix of linearized system of (1.3) about the fixed point $(\bar{x}, \bar{y})$ is given by

$$F_J(\bar{x}, \bar{y}) = \begin{bmatrix} C & -\frac{A\alpha\bar{y}^{\alpha-1}}{(1+\bar{y}^\alpha)^2} \\ -\frac{B\beta\bar{x}^{\beta-1}}{(1+\bar{x}^\beta)^2} & C \end{bmatrix}.$$

The characteristic polynomial of $F_J(\bar{x}, \bar{y})$ is given by

$$P(\lambda) = \lambda^2 - 2C\lambda + C^2 - \frac{AB\bar{x}^{\beta-1}\bar{y}^{\alpha-1}\alpha\beta}{(1+\bar{x}^\beta)^2(1+\bar{y}^\alpha)^2}. \tag{2.3}$$

From (2.1) it follows that

$$\bar{x} = \frac{A}{(1-C)(1+\bar{y}^\alpha)} \ , \ \bar{y} = \frac{B}{(1-C)(1+\bar{x}^\beta)}. \tag{2.4}$$

Using relations (2.4) in (2.3), the characteristic polynomial of $F_J(\bar{x}, \bar{y})$ can be written as

$$P(\lambda) = \lambda^2 - 2C\lambda + C^2 - \frac{\alpha\beta(1-C)^4\bar{x}^{\beta+1}\bar{y}^{\alpha+1}}{AB}. \tag{2.5}$$

Let $\phi(\lambda) = \lambda^2$ and $\psi(\lambda) = 2C\lambda - C^2 + \frac{\alpha\beta(1-C)^4\bar{x}^{\beta+1}\bar{y}^{\alpha+1}}{AB}$. Furthermore, assume that $|\lambda| = 1$, then it follows that

$$|\psi(\lambda)| < 2C + C^2 + \frac{\alpha\beta(1-C)^4\bar{x}^{\beta+1}\bar{y}^{\alpha+1}}{AB}$$

$$< 2C + C^2 + \frac{\alpha\beta A^\beta B^\alpha}{(1-C)^{\alpha+\beta-2}} < 1.$$

Lemma 2.3 (Grove and Ladas, 2004): Supposes that $I_1 = [\alpha, \beta]$ and $I_2 = [\gamma, \delta]$ be intervals of real numbers, and assume that $f_1: I_1 \times I_2 \longrightarrow I_1$ and $f_2: I_1 \times I_2 \longrightarrow I_2$ are continuous functions. Assume the following system

$$x_{n+1} = f_1(x_n, y_n), y_{n+1} = f_2(x_n, y_n) \tag{2.6}$$

where initial conditions $(x_0, y_0) \in I_1 \times I_2$. Let the following conditions are true:

(i) $f_1(x, y)$ is non-decreasing in x and non-increasing in y.

(ii) $f_2(x, y)$ is non-increasing in x and non-decreasing in y.

(iii) If $(m_1, M_1, m_2, M_2) \in I_1^2 \times I_2^2$ be a solution of the system:

$$m_1 = f_1(m_1, M_2), \qquad M_1 = f_1(M_1, m_2)$$
$$m_2 = f_2(M_1, m_2), \quad M_2 = f_2(m_1, M_2)$$

such that $m_1 = M_1$ and $m_2 = M_2$. Then, there exists exactly one fixed point $(\bar{x}, \bar{y})$ of the system (2.6) such that $\lim_{n\to\infty}(x_n, y_n) = (\bar{x}, \bar{y})$.

Theorem 2.5: The unique positive equilibrium point of the system (1.3) is a global attractor, if the following condition is satisfied:

$$(1 - C)^{2\alpha}(1 + L_1^\alpha)(1 + L_2^\alpha)^2 > \alpha^2(AB)^\alpha, \tag{2.7}$$

where $L_1 = \frac{A}{1-C}\left(\frac{1}{1+\left(\frac{B}{1-C}\right)^\alpha}\right)$ and $L_2 = \frac{B}{1-C}\left(\frac{1}{1+\left(\frac{A}{1-C}\right)^\beta}\right)$.

Proof: Let $f_1(x, y) = \frac{A}{1+y^\alpha} + Cx$, and $f_2(x, y) = \frac{B}{1+x^\beta} + Cy$. Then it is simple to see that $f_1(x, y)$ is non-decreasing in x and non-increasing in y. Moreover, $f_2(x, y)$ is non-increasing in x and non-decreasing in y. Let (m_1, M_1, m_2, M_2) be a positive solution of the system

$$m_1 = f_1(m_1, M_2), \qquad M_1 = f_1(M_1, m_2)$$
$$m_2 = f_2(M_1, m_2), \quad M_2 = f_2(m_1, M_2)$$

Then,

$$m_1 = \frac{A}{1+M_2{}^\alpha} + Cm_1, \quad M_1 = \frac{A}{1+m_2{}^\alpha} + CM_1, \tag{2.8}$$

and

$$m_2 = \frac{B}{1+M_1{}^\beta} + Cm_2, \quad M_2 = \frac{B}{1+m_1{}^\beta} + CM_2. \tag{2.9}$$

Then it follows from (2.8) and (2.9) that

$$L_1 = \frac{A}{1-C}\left(\frac{1}{1+\left(\frac{B}{1-C}\right)^\alpha}\right) \leq m_1 \leq M_1 \leq \frac{A}{1-C}, \tag{2.10}$$

$$L_2 = \frac{B}{1-C}\left(\frac{1}{1+\left(\frac{A}{1-C}\right)^\beta}\right) \leq m_2 \leq M_2 \leq \frac{B}{1-C}. \tag{2.11}$$

On subtracting (2.8), we obtain

$$(1-C)(M_1 - m_1) = A\left(\frac{M_2{}^\alpha - m_2{}^\alpha}{(1+m_2{}^\alpha)(1+M_2{}^\alpha)}\right)$$

$$= A\alpha\theta^{\alpha-1}\frac{(M_2-m_2)}{(1+m_2{}^\alpha)(1+M_2{}^\alpha)}, \tag{2.12}$$

where $m_2 \leq \theta \leq M_2$. From (2.11) and (2.12), it follows that

$$M_1 - m_1 \leq \frac{\alpha A B^{\alpha-1}}{(1-C)^\alpha(1+L_2{}^\alpha)^2}(M_2 - m_2). \tag{2.13}$$

Similarly subtracting (2.9) and using (2.10), we have

$$M_2 - m_2 \leq \frac{\alpha B A^{\alpha-1}}{(1-C)^\alpha(1+L_1{}^\alpha)^2}(M_1 - m_1). \tag{2.14}$$

Finally, from (2.13) and (2.14), one has

$$[(1-C)^{2\alpha}(1+L_1{}^\alpha)(1+L_2{}^\alpha)^2 - \alpha^2(AB)^\alpha](M_1 - m_1) \leq 0. \tag{2.15}$$

Under the condition (2.7), it follows from (2.15) that $M_1 = m_1$. Similarly, one has $M_2 = m_2$.

Lemma 2.4: Under the conditions of Theorem 2.4 and Theorem 2.5 the unique positive equilibrium point of system (1.3) is globally asymptotically stable.

References

Ahmad S. 1993. On the nonautonomous Lotka-Volterra competition equation. Proceedings of American Mathematical Society, 117(1993): 199-204

Din Q. 2013. Dynamics of a discrete Lotka-Volterra model. Advances in Differential Equations, 1: 1-13

Din Q. 2014. Global stability of a population model, Chaos, Soliton and Fractals, 59: 119-128

Din Q, Donchev T. 2013. Global character of a host-parasite model, Chaos, Soliton and Fractals, 54: 1-7

Gardner TS, Cantor CR, Collins JJ. 2000. Construction of a genetic toggle switch in *Escherichia coli*. Nature, 403(20): 339-342

Grove EA, Ladas G. 2004. Periodicities in Nonlinear Difference Equations. Chapman and Hall/CRC Press, Boca Raton, USA

Liu X. 2010. A note on the existence of periodic solution in discrete predator-prey models. Applied Mathematical Modeling, 34: 2477-2483

Papaschinopoulos G, Radin MA, Schinas CJ. 2011. On the system of two difference equations of exponential form: $x_{n+1} = a + bx_{n-1}e^{-y_n}, \quad y_{n+1} = c + dy_{n-1}e^{-x_n}$. Mathematical and Computer Modeling, 54:

2969-2977

Sedaghat H. 2003. Nonlinear Difference Equations: Theory with Applications to Social Science Models. Kluwer Academic Publishers, Dordrecht, Netherlands

Tang X, Zou X. 2006. On positive periodic solutions of Lotka-Volterra competition systems with deviating arguments. Proceedings of the American Mathematical Society, 134: 2967-2974

Zhou Z, Zou X. 2003. Stable periodic solutions in a discrete periodic logistic equation. Applied Mathematics Letters, 16(2): 165-171

16

Some topological properties of arthropod food webs in paddy fields of South China

LiQin Jiang[1], WenJun Zhang[1,2], Xin Li[3]

[1]School of Life Sciences, Sun Yat-sen University, Guangzhou 510275, China; [2]International Academy of Ecology and Environmental Sciences, Hong Kong

[3]College of Plant Protection, Northwest A & F University, Yangling 712100, China; Yangling Institute of Modern Agricultural Standardization, Yangling 712100, China

E-mail: zhwj@mail.sysu.edu.cn, lixin57@hotmail.com

Abstract

To explore the topological properties of paddy arthropod food webs is of significance for understanding natural equilibrium of rice pests. In present study, we used Pajek software to analyze the topological properties of four full arthropod food webs in South China. The results showed that predators were significantly abundant than preys, and the proportion of predators to preys (3.07) was significantly higher than previously reported by Cohen in 1977 (1.33). In the food webs, the number of top species was the largest, accounted for about 50% of the total. The number of intermediate-intermediate links was far greater than the other three links. The average degree of paddy arthropod food webs is 6.0, 6.04, 5.74 and 7.75, respectively. Average degree and link density did not change significantly with the change of the number of species, but the connectance reduced significantly. In the paddy ecosystems, the increase of species diversity does not lead to an increase proportionally to the links among species. The link density and connectance of food webs of early season rice field were less than that from late season rice field. Cycles of all food webs cycles were 0. The maximum chain length of the basal species was 3, and the largest chain length of the top species was typically 2 or 3. Neutral insects were found to play a very important role in the paddy ecosystem. *Nilaparvata lugens* and *Sogatella furcifera* were found to be the dominant species of rice pests. *Pardosa pseudoannulata, Tetragnatha maxillosa, Pirata subparaticus, Arctosa stigmosa* and *Clubiona corrugate* were identified as the important predatory species that may effectively control the pest population. The keystone species calculated from keystone index and network analysis are analogous, indicating either keystone index or network analysis can be used in the analysis of keystone species.

Keywords food webs; topological properties; paddy ecosystems; arthropods; natural enemies.

1 Introduction

Arthropod food webs in paddy ecosystems are complex ecological networks, which primarily describe the relationship between natural enemies and rice pests. A food web can explicitly express between-species trophic relationship. Studies on food webs can provide new ideas for rice pest control and management (Valladares et al., 1999). Arthropods are one of the most important organisms in paddy ecosystems (Zhang, 2011). The changes of a paddy arthropod community may indicate the occurrence situation and development trends of rice pests. Therefore the research on paddy arthropod food webs is one of the fundamental works to optimize the biological or natural control of rice pests (You et al., 1993).

Trophic relationship between species within a biome are expressed as the directional links between species in the food web, which depicts the intrinsic attributes of interdependence, mutual restrain and co-evolutionary relationship between the various organisms. Food webs are important part of the studies of biological communities (Price, 1981; Crichlow et al., 1982; DeAngelis et al., 1989; Zhang, 2007, 2011, 2012a, 2012b, 2012c). In recent years, a lot of studies have been done on food webs, including arthropod food webs (Guo et al., 1995; Crook et al., 1984; Prabhakar et al., 2012). For example, Jiang et al. (2006) recorded the dynamics of arthropod communities in paddy fields of Anhui Province.Yuan et al. (2010) studied the community structure of organic rice fields in Yangtze River farms and evaluated the effects of natural enemies on the control of rice insect pests. Wang et al. (2013) compared the community structure of arthropods between ecological and conventional rice fields. Furthermore, studies have indicated that climatic conditions (Zhang et al., 1997), pesticides, pest-resistance varieties of rice, and water-saving irrigation (Fuet al., 2013) would affect paddy arthropod communities. Overall most studies have focused on the effects of changes in ecosystems in different habitats, such as environmental changes, invasive species, species extinction, etc., on the structural components and dynamics of arthropod food webs. For example, Oraze (1988) studied the changes of spider community in the flooded rice fields. Gratton and Denno (2006) restored an arthropod food web following removal of an invasive plant. The concept of neutral insects was first put forward by Wu (1994). Neutral insects are defined as the insects neither natural enemies nor insect pests in rice ecosystems, such as chironomids, mosquitoes, flies and springtails, etc (Guo et al., 1995; Liu, 2000; Liu et al., 2002). Meanwhile, many researchers used different methods, such as serological method (Crook et al., 1984), population dynamics investigation method, ELISA method (Zhang et al., 1996; Liu et al., 2002), and the isotope method (Schmidt SN et al., 2007)to study the relationship between natural enemies and neutral insects and insect pests,, in order to guide utilization and protection of natural enemies and neutral insects.

The topological properties of food webs have been a hot topic since the presentation of food web concept (MacArther, 1955; Sprules and Bowerman, 1988; Hall and Raffaelli, 1991; Lafferty et al., 2006; Rzanny and Voigt, 2012).Some basic properties of food webs, including the number of species, the number of links, connectance, link density and the relationship among them were studied (Sugihara et al., 1989; Dunne et al., 2002; Navia et al., 2010). These properties stressed the importance of species in maintaining the stability of food webs. Nevertheless, so far the research on the topological properties of paddy arthropod food webs is fewer. In present study, we analyzed some topological properties of paddy arthropod food webs, aiming to provide a theoretical basis for improving the structure of arthropod food webs and for protecting natural enemies of insect pests.

2 Materials and Methods

2.1 Materials

2.1.1 Data sources

The data sources are listed as follows:

Name	Matrix (S×S)	District	Period	Data sources
FW1	26×26	Dasha Guangdong	The overall pattern	Liu et al. (2002)
FW2	57×57	Hunan	The overall pattern	Liu (2009)
FW3a	23×23	Wengyuan, Guangdong	The overall pattern of early season rice	Gu et al.(2006)
FW3b	24×24	Wengyuan, Guangdong	The overall pattern of late season rice	Gu et al.(2006)

2.1.2 Data description

Paddy arthropod food webs are composed of natural enemies, pests, neutral insects and plants. The food webs in present study primarily describe the relationship between the natural enemies and rice pests. The arthropod food web FW1 contains 24 species of arthropods, including 19 species of predators and 5species of preys which contain 4 rice insect pest species and 1 neutral insect species (Table 1). There are 55 arthropod species in FW2, including 36 predator species, and 19 prey species which contain 13 rice insect pest species and 6 neutral insect species (Table 2). FW3a has 21 arthropod species, including 16 species of predators, and 5 prey species which contain 4 rice insect pest species and 1 neutral insect species (Table 3). FW3b contains 22 arthropod species, including 17 predator species and 5 prey species in which there are 4 rice insect pest species and 1 neutral insect species (Table 4).

Table 1 Species and their roles in FW1.

ID	Species	Category	ID	Species	Category
1	*Araneus inustus*	predator	14	*Marpissa magister*	predator
2	*Dyschiriognatha quadrimaculata*	predator	15	*Microvelia horvathi*	predator
3	*Tetragnatha nitens*	predator	16	*Cyrtorrhinus livdipennis Reuter*	predator
4	*Coleosoma octomaculatum*	predator	17	*Casnoidea indica*	predator
5	*Hylyphantes graminicola*	predator	18	*Paederus fuscipesCurti*	predator
6	*Ummeliata insecticeps*	predator	19	*Micraspis discolor*	predator
7	*Pirata subparaticus*	predator	20	*Cnaphalocrocis medinalis Guenee*	prey
8	*Pardosa pseudoannulata*	predator	21	*Sogatella furcifera*	prey
9	*Pardosa tschekiangensis*	predator	22	*Nilaparvata lugens*	prey
10	*Clubiona corrugata*	predator	23	*Oxya chinensis*	prey
11	*Clubiona corrugata*	predator	24	*Chironomus sp*	prey
12	*Oxyopes sertatus*	predator	25	*Rice*	
13	*Bianor hotingchiechi*	predator	26	*Humus*	

Table 2 Species and their roles in FW2.

ID	Species	Category	ID	Species	Category
1	*Pirata japonious*	predator	29	*Paederus fuscipes Curti*	predator
2	*Pirata subparaticus*	predator	30	*Micraspis discolor*	predator
3	*Pardosa pseudoannulata*	predator	31	*Casnoidea indica*	predator
4	*Pardosa tschekiangensis*	predator	32	*Ophionea indica*	predator
5	*Tetragnatha nitens*	predator	33	*Colliuris chaudoiri Bohem*	predator
6	*Neoscona nautica*	predator	34	*Carabiade*	predator
7	*Neoscona theisi*	predator	35	*Coccinella septempunctata*	predator
8	*Neoscona griseomaculata*	predator	36	*Harmonia axyridis*	predator
9	*Acusilas coccneus*	predator	37	*Culex triaeniorhynchus*	prey
10	*Araneidae*	predator	38	*Chironomus sp*	prey
11	*Araneus inustus*	predator	39	*Salina sp*	prey
12	*Argiope aemula*	predator	40	*Entomobrya griseoolivata*	prey
13	*Dyschiriognatha quadrimaculata*	predator	41	*Hypogastramatura*	prey
14	*Coleosoma octomaculatum*	predator	42	*Bourletiella christianseni*	prey
15	*Clubiona corrugata*	predator	43	*Sogatella furcifera*	prey
16	*Bianor hotingchiechi*	predator	44	*Nilaparvata lugens*	prey
17	*Salticidae*	predator	45	*Oxya chinensis*	prey
18	*Ummeliata insecticeps*	predator	46	*Naranga aenesc*	prey
19	*Hylyphantes graminicola*	predator	47	*Cnaphalocrocis medinalis Guenee*	prey
20	*Oxyopes sertatus*	predator	48	*Tryporyza incertulas*	prey
21	*Ebrechtella tricuspidata*	predator	49	*Nephotettix bipunctatus*	prey
22	*Marpissa magister*	predator	50	*Empoasea subrufa*	prey
23	*Clubiona corrugata*	predator	51	*Tettigoniella spectra*	prey
24	*Tetragnatha maxillasa*	predator	52	*Mythimna separata*	prey
25	*Dolomedes sp*	predator	53	*Tettigoniidae*	prey
26	*Plecippussetipe sp*	predator	54	*Oxya chinensis*	prey
27	*Cyrtorrhinus livdipennis Reuter*	predator	55	*Nephotettix cincticeps*	prey
28	*Microvelia horvathi*	predator	56	*Rice*	
			57	*Humus*	

Table 3 Species and their roles in FW3a.

ID	Specifies	Category	ID	Species	Category
1	*Pirata piratoides*	predator	13	*Ummeliata insecticeps*	predator
2	*Tetragnatha maxillosa*	predator	14	*Pardosa pseudoannulata*	predator
3	*Neoscona nautica*	predator	15	*Pirata subparaticus*	predator
4	*Tetragnatha nitens*	predator	16	*Bianor aurocinctus*	predator
5	*Tetragnatha mandibulata*	predator	17	*Sogatella furcifera*	prey
6	*Thalassius affinis*	predator	18	*Nilaparvata lugens*	prey
7	*Micraspis discolor*	predator	19	*Cnaphalocrocis medinalis Guenee*	prey
8	*Staphylinidae*	predator	20	*Oxya chinensis*	prey

9	*Cyrtorrhinus livdipennis Reuter*	predator	21	*Chironomus sp*		prey
10	*Oxyopes lineatipes*	predator	22	*Rice*		
11	*Bianor hotingchiehi Schenke*	predator	23	*Humus*		
12	*Pardosa laura*	predator				

Table 4 Species and their roles in FW3b.

ID	Species	Category	ID	Species	Category
1	*Arctosa stigmosa*	predator	13	*Ummeliata insecticeps*	predator
2	*Tetragnatha maxillosa*	predator	14	*Pardosa pseudoannulata*	predator
3	*Neoscona nautica*	predator	15	*Pirata subparaticus*	predator
4	*Tetragnatha nitens*	predator	16	*Pardosa laura*	predator
5	*Tetragnatha mandibulata*	predator	17	*Bianor aurocinctus*	predator
6	*Tetragnatha caudicula*	predator	18	*Sogatella furcifera*	prey
7	*Micraspis discolor*	predator	19	*Nilaparvata lugens*	prey
8	*Staphylinidae*	predator	20	*Cnaphalocrocis medinalis Guenee*	prey
9	*Cyrtorrhinus livdipennis Reuter*	predator	21	*Oxya chinensis*	prey
10	*Oxyopes lineatipes*	predator	22	*Chironomus sp*	prey
11	*Leucauge blanda*	predator	23	*Rice*	
12	*Coleosoma octomaculatum*	predator	24	*Humus*	

2.1.3 Data conversion

Species were labeled with ID codes (see Tables 1, 2, 3 and 4). In the Pajek environment, choose the directory and execute the command as follows: Open data →Data editors →Matrix editor, in the UCINET software, and save them as the files in ".##h" format. Finally, choose the directory and execute the command: File→Open→Ucinet dataset→network, in Netdraw software; choose and open the ".##h" file, and then save it to the file in ".net" format by the command: File→Save data as→Pajek→Net file.The resultant four".net" files formed the basis for topological analysis.

2.2 Methods

2.2.1 Pajek software (Network analysis)

Pajek is a software platform for the network analysis of the large and complex networks with up to millions of nodes. It is a fast visualized tool for program operation. Pajek contains various methods/algorithms on analysis of topological properties.

2.2.2 Topological properties and measures

2.2.2.1 Classification of species

Species in a food web can be divided into three categories, top species T, intermediate species I, and basal species B (Pimm et al., 1991). Atop species is a species not eaten by any species in the web. An intermediate species is a species that has both at least one predator and at least one prey. A basal species is a species that eats no species.

2.2.2.2 Link analysis

Links in a food web can be divided into four categories, the basal-intermediate links, the basal-top links, the intermediate-intermediate links, and the intermediate-the top links (Cohen and Newman, 1985). For example, a basal-intermediate link is a link from a basal species to an intermediate species.

2.2.2.3 Degree analysis

Degree is a basic property for a network. The degree of a node is defined as the number of its connected nodes. In general, the greater the degree of a node, the more important the node is in the food web (Zhang, 2012d). We obtained the degree of nodes by performing the command: Net→Parations→Degree→In/Out/All in Pajek software, where All is the sum of outgoing degree and incoming degree.

2.2.2.4 Connectance and link density

Connectance is defined as the number of observed trophic interactions divided by the number of possible interactions (Zhang, 2012a, 2012d). The number of possible interactions may be S^2 if cannibalistic interactions are counted, and $S(S-1)$ if only interspecific interactions are counted. Link density is equal to the ratio of total number of links to the total number of species.

2.2.2.5 Chain cycle analysis

A chain cycle refers to a closed loop in the food chain. For example, cannibalism is a cycle where one species feeds upon itself. In Pajek, chain cycles can be obtained by using command: Net→Count→4-rings→directed→cyclic.

2.2.2.6 Chain length analysis

Chain length is defined as the number of links connected to each other through two adjacent species between the basal species and the top species. We obtained chain length by the command: Net → K-neigbours→input/output.

2.2.3 Keystone index

Keystone index is a two-way trapezoidal index, and proposed by Jordán et al. (1999) based on the food web. It includes top-down and bottom-up control of material flow and information flow in food webs, namely K_b for botton-up, K_t for top-down and K for bidirectional processes ($K = K_b + K_t$). The specific formula is as follows:

$$K_b(i) = \frac{1 + K_b(j)}{m(i)(j)}$$

$$K_t(i) = \frac{1 + K_t(j)}{n(i)(j)}$$

$$K(i) = K_b(i) + K_t(i)$$

where $K_b(j)$ is the bottom-up keystone index of the jth predator, $m(i)(j)$ is the number of its direct preys. $K_t(j)$ is the top-down keystone index of the jth prey, and $n(i)(j)$ is the number of its direct predators. Keystone index is a measure on the basis of topological structure of food web. Thus it is theoretically similar to some measures in network analysis.

3 Results

3.1 Species analysis

As indicated in Tables 1, 2, 3 and 4, the numbers of predators/preys in the four arthropod food webs are 19/5, 36/19, 16/5 and 17/5, respectively. The average number of each predator feeds on prey species is 3.8, 1.9, 3.2 and 3.4, respectively. This is basically different from that of Cohen (1977) (4:3, i.e., 1.33). It is found that the more species in the food web, the less average number of each predator feeds on prey.

Briand and Cohen (1984) proposed that top species, intermediate species and basal species were all approximately proportional to the number of total species, and the proportion was 0.29, 0.53 and 0.19, respectively. Table 5 exhibits that in paddy arthropod food webs, the number of predators is significantly greater than that of preys. The proportion of top species is the largest, with about half of the total species, and the proportion of basal species is the least.

Table 5 Species analysis of food webs.

Food web	Trophic level	Number of species	Total number of species	Proportion	Species ID
FW1	T	13	26	50%	1-6, 8-11, 13-15
	I	11		42.3%	7, 12, 16-24
	B	2		7.7%	25,26
FW2	T	31	57	54.4%	1, 3-19, 21-26, 28, 31-36
	I	24		42.11%	2, 20, 27, 29, 30, 37-55
	B	2		3.49%	56, 57
FW3a	T	11	23	47.83%	1-6, 11-13, 15, 16
	I	10		43.48%	7-10, 14, 17-21
	B	2		8.69%	22,23
FW3b	T	12	24	50%	1-6, 11-14, 16, 17
	I	10		41.67%	7-10, 15, 18-22
	B	2		8.33%	23, 24

Table 6 Link analysis of food webs.

Food web	Trophic level	Number of links	Total number of links	Proportion
FW1	B-I	5	78	6.41%
	B-T	0		0
	I-I	21		26.92%
	I-T	52		66.67%
FW2	B-I	19	172	11.04%
	B-T	0		0
	I-I	32		18.61%
	I-T	121		70.35%
FW3a	B-I	5	66	7.57%
	B-T	0		0
	I-I	11		16.67%
	I-T	50		75.76%
FW3b	B-I	5	93	5.38%
	B-T	0		0
	I-I	14		15.05%
	I-T	74		79.57%

3.2 Link analysis

Briand and Cohen (1984) proposed that the proportion T: I: B of food webs is a constant. Similarly, the proportions of the basal-intermediate links, the basal-top links, the intermediate-intermediate links, and the intermediate-top links are also constants (0.27, 0.08, 0.30 and 0.35, respectively). However, the link analyses of paddy arthropod food webs showed that the number of the intermediate-intermediate links is far greater than the number of the other three types of links, and the number of the basal-top links in all food webs is 0.

Therefore, the link density and connectance of FW1 are3 and 0.12, respectively; for FW2 they are 3.02 and 0.05, respectively; for FW3a they are 2.87 and 0.13, respectively, and for FW3b they are 3.88 and 0.17, respectively. Obviously, the link density of all food webs is similar to each other. But FW2, which harbors the richest species, has the smallest connectance. It means that in the paddy ecosystems, the increase of species diversity does not lead to an increase proportionally to the links among species. In addition, the link density and connectance of early season rice field are less than that of late season rice field. This demonstrates that the number of links in late season rice fieldis richer than that in the early rich field, i.e., the interactions between predators and preys in the late season rice field are more active than that in the early season rice field.

3.3 Analysis of chain cycle and chain length

Cycles of all paddy arthropod food webs are zeros, which is in consistent with the conclusion that the cycles are rare in the food webs (Pimm et al., 1991).

Chain length analysis of species No. 25 and 26 in FW1 indicates that the chain length of rice and humus is 3 (Fig. 1). Similarly, chain length of FW2, FW3a and FW3b is 3.

(a)   (b)

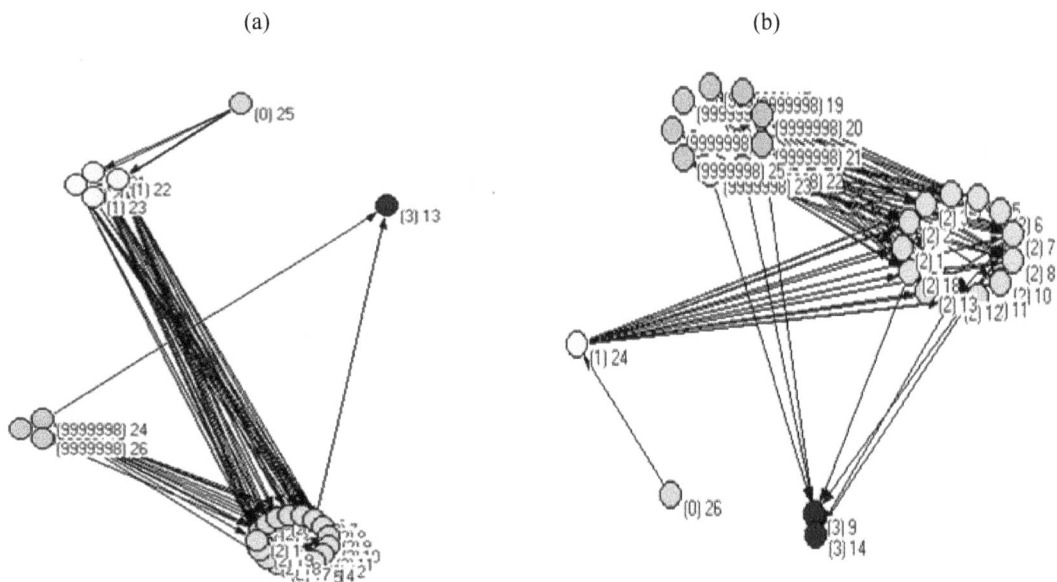

Fig. 1 Chain length analysis of species No. 25 (a) and 26 (b) in FW1. For each species, the number in parenthesis is chain length and the number outside parenthesis is species ID code. The species No. 9,999,998 means that it is not reachable to the current node.

Chain length analysis of top species demonstrates that the maximum chain length from the basal species to the top species is typically 2 or 3; chain length 1 is rarely found, and the chain length larger than 3 occurs seldom (Table 7), which are in consistent with the conclusions of Pimm et al. (1991).

Table 7 Chain length analysis of top species in food webs.

Top species (ID) in FW1	1	2	3	4	5	6	8	9	10	11	13	14	15			
Maximum chain length	2	3	2	2	2	2	2	3	2	2	3	2	2			
Top species (ID) in FW2	1	3	4	5-12	13	14	15	16	17	18	19	21	22	23-26	28	31-36
Maximum chain length	2	2	3	2	3	2	2	4	2	2	2	2	2	2	2	2
Top species (ID) in FW3a	1	2	3	4	5	6	11	12	13	15	16					
Maximum chain length	3	2	2	2	2	2	3	2	2	2	1					
Top species (ID) in FW3b	1	2	3	4	5	6	11	12	13	14	16	17				
Maximum chain length	2	2	2	2	2	3	2	2	2	2	1	3				

3.4 identification analysis of keystone species

By Pajek analysis, the average degrees of paddy arthropod food webs are 6.0, 6.04, 5.74 and 7.75, respectively. As indicated in Fig.2, in FW1 and FW2, the species with the largest incoming degree is *Pardosa pseudoannulata*, followed by *Pirata subparaticus* and *Clubiona corrugate*; namely *Pardosa pseudoannulata* has the widercontrol spectrum on the rice pests. In FW3a, both *Pardosa pseudoannulata* and *Tetragnatha maxillosa*are the most significant. The most significant species are *Pardosa pseudoannulata*, *Tetragnatha maxillosa* and *Arctosa stigmosa*in FW3b.

(a)   (b)

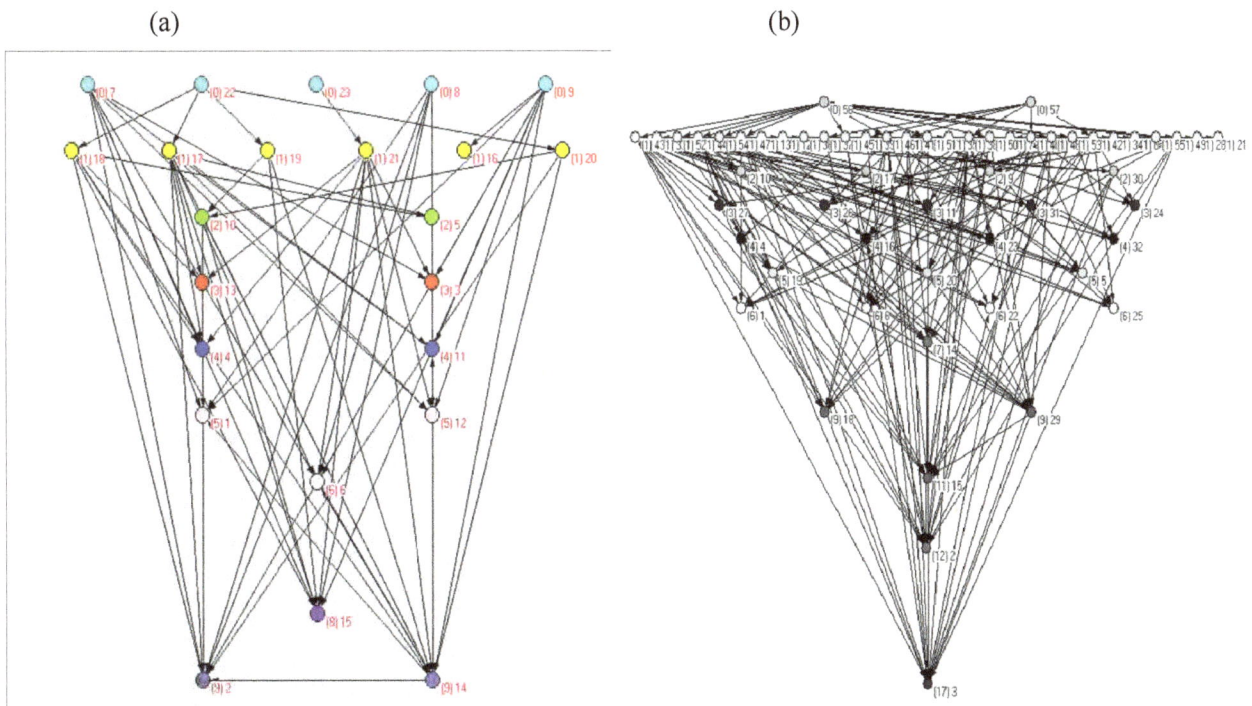

(c)   (d)

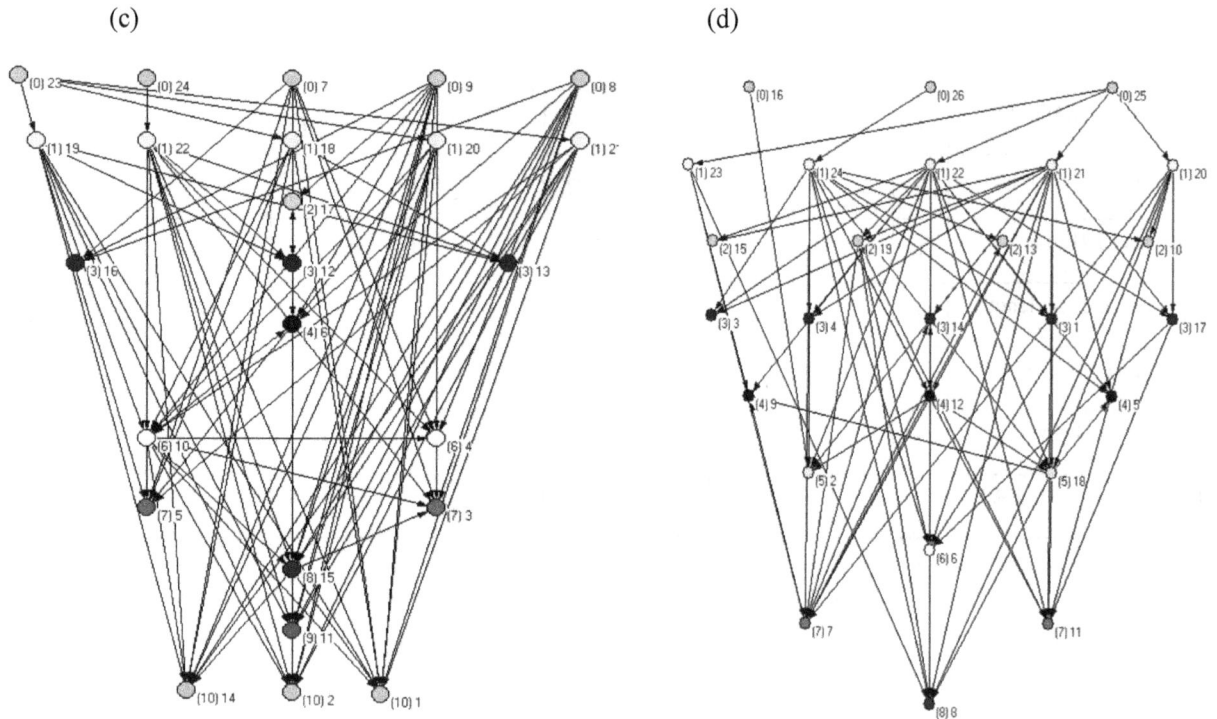

Fig. 2 Food web links of incoming degree analysis onfood webs FW1 (a), FW2 (b), FW3a(c) and FW3b (d), respectively. For each species, the number in parenthesis is incoming degree and the number outside parenthesis is species ID code.

Gu et al. (2006) argued that the arthropod species *Pardosa pseudoannulata, Ummeliata insecticeps, Pirata subparaticus* and *Tetragnatha maxillosa* would play an important role in controlling insect pests in paddy ecosystems, among which *Pardosa pseudoannulata* and *Ummeliata insecticeps* were the dominant species. In present network study, we further confirmed that *Pardosa pseudoannulata* held an absolutely important position in paddy arthropod food webs. However, *Ummeliata insecticeps* did not show its importance in food webs. This may be attributed to that *Ummeliata insecticeps* doesn't feed on other predators and not be preyed by other predators;it only feeds on pests and neutral insects, leading to its lower incoming degree than other predators.The dominant speciesinFW3a and FW3b are similar to each other, but *Arctosa stigmosa*did not proposed as a dominant species in Gu et al. (2006). In summary, *Pardosa pseudoannulata, Tetragnatha maxillosa, Pirata subparaticus, Clubiona corrugate* and *Arctosa stigmosa* are important species in paddy arthropod food webs.

From species composition of different food webs, we conclude that the important species in the paddy arthropod communities would not be relevant to the number of prey species.

However, the results as shown in Table 8 through the top-down keystone index, K_t, the species with the largest K_t value in FW1 is *Pardosa pseudoannulata*, followed by *Clubiona corrugate* and *Pardosa tschekiangensis*; The species has the widest control spectrum on the rice pests in FW2 is *Pardosa pseudoannulata*, followed by *Clubiona corrugate* and *Pirata subparaticus*; In FW3a, *Tetragnatha maxillosa* is the most significant species, followed by *Pardosa pseudoannulata* and *Pirata subparaticus*; In FW3b, The most significant species are *Arctosa stigmosa, Tetragnatha maxillosa, Pardosa pseudoannulata*. Compared with network analysis, the key species obtained by ecological analysis did not produce much difference. The key species in FW1 and FW2 is *Pardosa pseudoannulata*, but *Pirata subparaticus* in FW1 did not show the critical importance in keystone species analysis, but *Pardosa tschekiangensis*. Furthermore, the keystone index results of FW3a and FW3b were consistent with the results of network analysis, but there were differences in their values, shown that the arrangement of the importance of species was different. Similarly, the most significant species *Ummeliata insecticeps* in Gu et al. (2006) did not show a larger K_t value.

Table 8 Species with higher K_t value and their ID codes in the four paddy arthropod food webs.

	FW1			FW2			FW3a			FW3b	
ID	K_t value	species	ID	K_t value	species	ID	K_t value	species	ID	K_t value	species
8	1.91	*Pardosa pseudoannulata*	3	4.71	*Pardosa pseudoannulata*	2	2.68	*Tetragnatha maxillosa*	1	1.7	*Arctosa stigmosa*
11	1.91	*Clubiona corrugata*	15	3.17	*Clubionacorrugata*	14	2.28	*Pardosa pseudoannulata*	2	1.7	*Tetragnatha maxillosa*
9	1.76	*Pardosa tschekiangensis*	2	3.13	*Pirata subparaticus*	15	1.86	*Pirata subparaticus*	14	1.7	*Pardosa pseudoannulata*
2	1.58	*Dyschiriognatha quadrimaculata*	4	2.37	*Pardosa tschekiangensis*	1	1.67	*Pirata piratoides*	11	1.46	*Leucauge blanda*
6	1.47	*Ummeliata insecticeps*	22	2.22	*Marpissa magister*	6	1.6	*Thalassius affinis*	3	1.22	*Neoscona nautica*
7	1.47	*Pirata subparaticus*	18	2.09	*Ummeliata insecticeps*	11	1.18	*Bianor hotingchiehi Schenke*	15	1.18	*Pirata subparaticus*
14	1.17	*Marpissa magister*	16	1.73	*Bianor hotingchiechi*	21	1	*Chironomus sp*	5	1.14	*Tetragnatha mandibulata*
24	1	*Chironomus sp*	29	1.29	*Paederus fuscipes Curti*	12	0.78	*Pardosa laura*	22	1	*Chironomus sp*

(a)

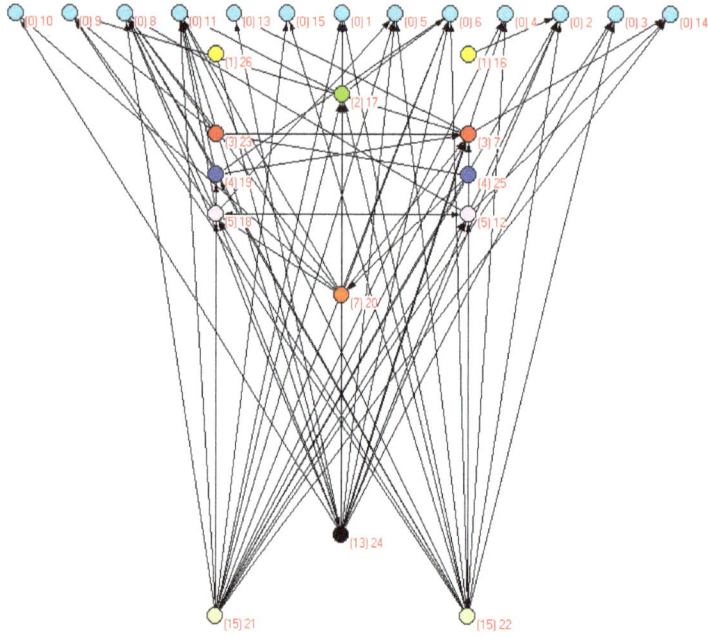

(b)

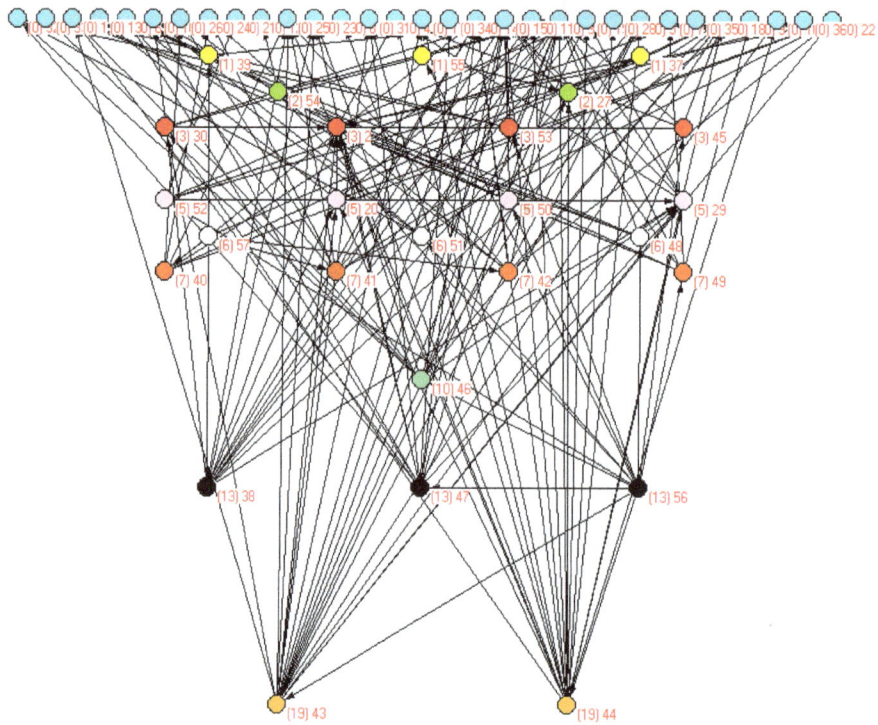

(c)

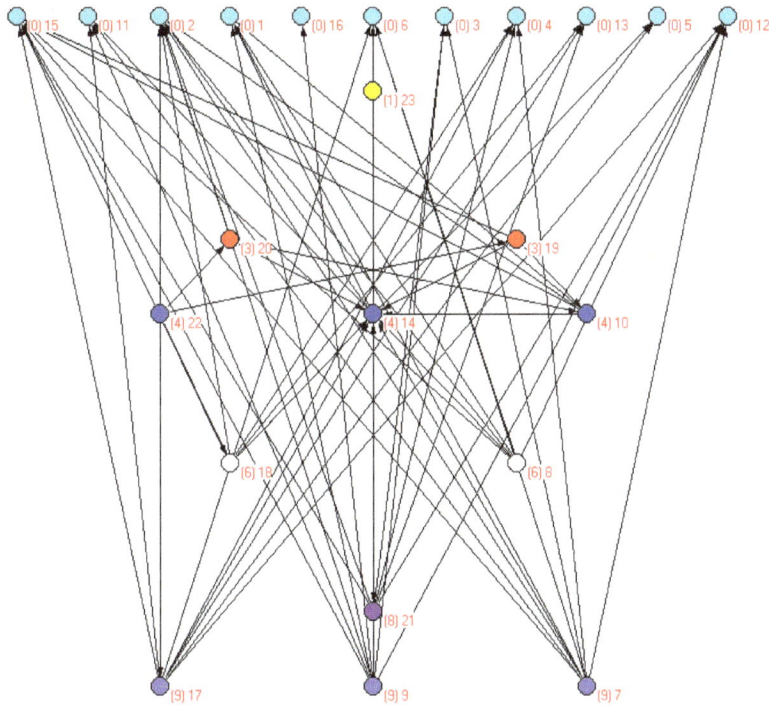

(d)

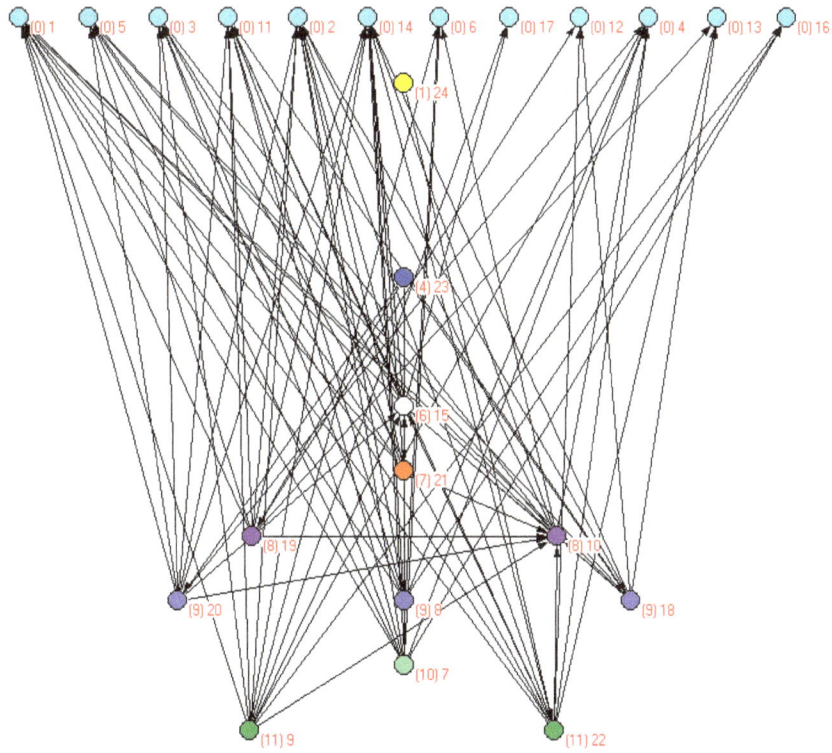

Fig. 3 Food web links of outgoing degree analysis of FW1 (a), FW2 (b), FW3a(c) and FW3b (d), respectively. For each species, the number in parenthesis is outgoing degree and the number outside parenthesis is species ID code.

It can be found from Fig. 3, that the outgoing degree of *Chironomus sp* is just less than *Nilaparvata lugens* and *Sogatella furcifera*, two rice pests in the FW1 and FW2. The outgoing degree of *Chironomussp* in FW3a is less than *Micraspis discolor, Cyrtorrhinus livdipennis Reuter* and *Sogatella furcifera*. In FW3b, *Chironomus sp* and *Cyrtorrhinus livdipennis Reuter* have the largest outgoing degree. In addition, the bottom-up keystone index, K_b, has shown in Table 9. In FW1 and FW2, both *Sogatella furcifera* and *Nilaparvata lugens* have the largest K_b values, and species with the largest K_b value in FW3a and FW3b is *Cyrtorrhinus livdipennis Reuter*. These results mean that *Nilaparvata lugens* and *Sogatella furcifera* are the dominant pest species, and their natural enemies are abundant also. *Cyrtorrhinus livdipennis Reuter* is the most significant predator. The outgoing degree of neutral insect, e.g., *Chironomussp*, further verifies its role as a complementary food in the arthropod food webs; it can be used as supplementary prey source of natural enemies.

Table 9 Species with higher K_b value and their ID codes in the four paddy arthropod food webs.

	FW1			FW2			FW3a			FW3b	
ID	K_b value	species	ID	K_b value	Species	ID	K_b value	species	ID	K_b value	species
21	5.43	*Sogatella furcifera*	44	6.01	*Nilaparvata lugens*	9	2.58	*Cyrtorrhinusli vdipennis Reuter*	9	2.44	*Cyrtorrhinus livdipennis Reuter*
22	5.35	*Nilaparvata lugens*	43	5.09	*Sogatella furcifera*	17	2.13	*Sogatella furcifera*	22	2.16	*Chironomus sp*
24	4.31	*Chironomus sp*	52	3.38	*Mythimnase parata*	7	1.83	*Micraspis discolor*	18	1.85	*Sogatella furcifera*
20	2	*Cnaphalocrocis medinalis Guenee*	47	3.23	*Cnaphalocrocis medinalis Guenee*	21	1.71	*Chironomus sp*	7	1.7	*Micraspis discolor*
12	1.31	*Oxyopes sertatus*	38	2.63	*Chironomus sp*	18	1.51	*Nilaparvata lugens*	19	1.6	*Nilaparvata lugens*

Comprehensive analysis of the bottom-up and top-down keystone index have shown in Table 10. K is sum of K_b and K_t. *Nilaparvata lugens* and *Sogatella furcifera* as rice pests in FW1 and FW2 have the largest K value, and the main keystone species in FW1 and FW2 is *Pardosa pseudoannulata, Pirata subparaticus, Clubiona corrugate* and *Pardosa tschekiangensis*. *Pardosa pseudoannulata, Tetragnatha maxillosa, Cyrtorrhinus livdipennis Reuter* and *Pirata subparaticus* in FW3a are the keystone species, and in FW3b, *Arctosa stigmosa* and *Oxyopes lineatipes* are also the keystone species. Spiders, as the major arthropod species, have significant control effect on the rice insect pests. Therefore, in order to achieve sustainable integrated pest prevention, it should strengthen the ecological protection of paddy field spiders.

Table 10 Species with higher *K* value and their ID codes in the four paddy arthropod food webs.

	FW1			FW2			FW3a			FW3b	
ID	*K* value	species	ID	*K* value	species	ID	*K* value	species	ID	*K* value	species
21	5.68	*Sogatella furcifera*	44	6.09	*Nilaparvata lugens*	14	3.01	*Pardosa pseudoannulata*	22	3.16	*Chironomus sp*
22	5.6	*Nilaparvata lugens*	43	5.17	*Sogatella furcifera*	21	2.71	*Chironomus sp*	9	2.44	*Cyrtorrhinus livdipennis Reuter*
24	5.31	*Chironomussp*	3	4.71	*Pardosa pseudoannulata*	2	2.68	*Tetragnatha maxillosa*	15	2.23	*Pirata subparaticus*
7	2.55	*Pirata subparaticus*	2	3.8	*Pirata subparaticus*	9	2.58	*Cyrtorrhinus livdipennis Reuter*	10	2.15	*Oxyopes lineatipes*
20	2.25	*Cnaphalocrocis medinalis Guenee*	52	3.46	*Mythimnase parata*	17	2.38	*Sogatella furcifera*	18	2.1	*Sogatella furcifera*
8	1.91	*Pardosa pseudoannulata*	47	3.31	*Cnaphalocrocis medinalis Guenee*	15	2.86	*Pirata subparaticus*	19	1.85	*Nilaparvata lugens*
11	1.91	*Clubiona corrugata*	15	3.17	*Clubiona corrugata*	7	1.83	*Micraspis discolor*	20	1.75	*Cnaphalocrocis medinalis Guenee*
9	1.76	*Pardosatschekiang ensis*	38	2.8	*Chironomus sp*	18	1.76	*Nilaparvata lugens*	1	1.7	*Arctosa stigmosa*
12	1.63	*Oxyopessertatus*	46	2.44	*Naranga aenesc*	1	1.67	*Pirata piratoides*	2	1.7	*Tetragnatha maxillosa*
2	1.58	*Dyschiriognathaqu adrimaculata*	4	2.37	*Pardosa tschekiangensis*	6	1.6	*Thalassius affinis*	7	1.7	*Pardosa pseudoannulata*

4 Conclusions and Discussion

Paddy arthropod food webs are generally complex ecosystems. Various ecological interactions can be found, including parasitism, predation, etc., which closely relate to biological control of rice insect pests. Therefore, the topological properties analysis of paddy arthropod food webs is a fundamental work for the biological control of insect pests. From analysis above, we draw some major conclusions as follows:

(1) Overall the ratios of predators to preys in the arthropod food webs are quite different from the results of Cohen (1977). The reason of the ratio for FW2 being close to the proposed by Cohen (1977) may be attributed to that the data of FW1, FW3a andFW3b were collected in the empirical fields, and the food web data of

Cohen and FW2weremostly qualitative collated. Under certain conditions, the number of species in a qualitative summary of food webs is significantly larger than the actual number of species found in the empirical fields. Therefore, a systematic review and analysis of paddy arthropod food webs should focus on practical (observed) communities.

(2)   Proportions of different trophic levels and the ratios of link types are different from Briand and Cohen (1984). In present study, the number of top species is about half of the total number of species, and the number of intermediate species is slightly little than the number of top species. The number of intermediate-intermediate links is far greater than the number of the other three kinds of links, and the basal-top links are all absent. These may be due to the absence of predators fed on predatory spiders.

(3)   Average degree and link density of arthropod food webs do not change much with the change of the number of species, but the connectance significantly reduces. Link density and connectance of the early season rice field and late season rice field show certain difference. Therefore, food webs should not be constructed through qualitative summary.

(4)   There are not cycles in arthropod food webs. The maximum chain length of the basal species is 3, and the largest chain length of the top species is typically 2 or 3, which are in consistent with Pimm et al (1991). Thus the topological properties of paddy arthropod food webs are in coincident with the cascade model, which can be further validated in future studies.

(5)   In the paddy ecosystems studied, *Pardosa pseudoannulata*is the dominant natural enemy species. The natural enemies *Tetragnatha maxillosa*, *Pirata subparaticus*, *Arctosa stigmosa* and *Clubiona corrugate* have stronger control effects on pests also. Furthermore, the outgoing degree and K value of *Chironomus sp* indicates that neutral insects play an important role in the paddy ecosystems (Guo, 1995).

(6)   The keystone species calculated from keystone index and network analysis are analogous, indicating either keystone index or network analysis can be used in the analysis of keystone species.

In present study, paddy arthropod food webs were constructed based on the matrixes representing relationship between pests, predators and neutral insects. Parasites, predatory birds and other predators were not included in the food webs. In future studies, we suggest that: (1) Complete food webs should further include parasites and predatory birds, etc. (2) In food web analysis, some models, such as the cascade model (Cohen and Newman, 1985), the niche model (Williams and Martinez, 2000) and nested model (Cattin et al., 2004) may be fitted and analyzed. (3) Both temporal and spatial aspects of food webs should be considered in order to provide a better theoretical basis for biological control and ecosystem maintenance in paddy fields.

Acknowledgment

We are thankful to the support of the project, Discovery and Crucial Node Analysis of Important Biological Networks (2015.6-2020.6), from Yangling Institute of Modern Agricultural Standardization, China.

References

Briand F, Cohen JE. 1984. Community food webs have scale-invariant structure. Nature, 307(5948): 264-267

Cattin MF, Bersier LF, et al. 2004.Phylogenetic constraints and adaptation explain food-web structure. Nature, 427: 835-839

Cohen JE. 1977. Ratio of prey to predators in community food webs. Nature, 270: 165-167

Cohen JE, Newman CM. 1985. A stochastic-theory of community food web. I. Models and aggregated data. Philosophical Transactions of the Royal Society of London Series B, Biological Sciences, 224: 421-448

Crichlow RE, Stearns SC. 1982. The structure of food webs. American Naturalist, 120(4): 478-499

Crook NE, Sunderland KD. 1984. Detection of aphid remains in predatory insects and spiders by ELISA. Annals of Applied Biology, 105:413-422

DeAngelis DL, Bartell SM, Brenkert AL. 1989. Effects of nutrient recycling and food-chain length on resilience. American Naturalist, 134(5): 778-805

Dunne JA, Williams RJ, Martinez ND. 2002. Network structure and biodiversity loss in food webs: robustness increases with connectance. Ecology Letters, 5: 558-567

Fu HL, Luo YF, Peng SZ, et al. 2013. Effects of water-saving irrigation on diversity of arthropod communities in paddy fields. Water-saving Irrigation, 10: 14-20

Gratton C, Denno RF. 2006. Arthropod food web restoration following removal of an invasive wetland plant. Ecological Applications, 16:622-631

Gu DX, Xia Q, Zhang GR, et al. 2006. The paddy spider community structure and trophic relationships. In: Structure and Fucntions of Spider Communities in Paddy Fields of China (Wang HQ, ed). 136-141, Hunan Normal University Press, Changsha, China

Gu ZY, Han LJ, Wang Q, et al. 1996. Ecological factors on the resurgence of rice planthopper after applying synthetic pyrethroids and their control. Entomological Journal of East China, 5(2): 87-92

Guo YJ, Wang NY, Jiang JW, et al. 1995. Ecological significance of neutral insects as nutrient bridge for predators in irrigated rice arthropod community.Chinese Journal of Biological Control. 11(1): 5-9

Hall SJ, Raffaelli. 1991. Lessons from a species-rice web. Journal of Animal Ecology. 60(3): 823-841

Jiang JQ, Miu Y, Li GT, et al. 2006. Dynamics of insect pest and natural enemy communities in middle–season paddy fields in the Yangtze-Huaihe area. Acta Agriculture Universitatis Jiangxiensis, 28(3): 354-358

Jordán F,Takacs-Santa A, Molnar I. 1999. Are liability theoretical quest for key stones. Oikos, 86: 453-462

Lafferty KD, Hechinger RF, Shaw JC, et al. 2006. Food webs and parasites in a salt marsh ecosystem. In Disease Ecology: Community Structure and Pathogen Dynamics (Collinge S, Ray C, eds). 119-134, Oxford University Press, Oxford, UK

Liu WH. 2009. The construction of paddy predatory arthropod food web.Journal of Xiangtan Normal University (Natural Science Edition), 31(1): 56-59

Liu YF. 2000. The Structure Analysis of Arthropod Communities in Paddy Ecosystem. PhD Thesis. School of Life Sciences, Sun Yat-sen University, Guangzhou, China

Liu YF, Gu DX, Zhang GR, et al. 2002. The structure analysis of predatory food webs in paddy ecosystem. In: Entomology Innovation and Development Engineers. 207-213, China Science and Technology Press, Beijing, China

Liu YF, Gu DX, Zhang GR, et al. 2002. Enemy-linked immunosorbent assay used to explore the predation of *Chironomus sp.* (*Diptera: Chironomide*) by predators in paddy fields. Acta Ecologica Sinica, 22(10): 1699-1703

MacArthur R. 1955. Fluctuation of animal populations and a measure of community stability. Ecology, 36(3): 533-536

Martinez ND. 1992. Constant connectance in community food webs. The American Naturalist, 139(6): 1208-1218

Navia AF, et al. 2010. Topological analysis of the ecological importance of elasmo branch fishes: A food web study on the Gulf of Tortugas, Colombia. Ecological Modelling, 221(24): 2918-2926

Oraze MJ, Grigarick AA, et al. 1988. Spider fauna of flooded rice fields in Northern California. American Arachnological Society, 16(3): 331-337

Pimm SL. The Balance of Nature. University of Chicago Press, USA, 1991

Prabhakar CS, Sood P, Mehta PK. 2012. Pictorial keys for predominant Bactrocera and Dacus fruit flies (*Diptera: Tephritidae*) of north western Himalaya. Arthropods, 1(3):101-111

Price PW. Insect Ecology (Translated by the Department of Entomology, Department of Biology), Peking University, Beijing: People's Education Press, China, 1981

Rzanny M, Voigt W.2012. Complexity of multitrophic interactions in a grassland ecosystem depends on plant species diversity. Journal of Animal Ecology, 81: 614-627

Schmidt SN, Olden JD, Solomon CT, et al. 2007. Quantitative approaches to the analysis of stable isotope food web data. Ecology, 88(11): 2793-2802

Sprules WG,Bowerman JE. 1988. Omnivory and food chain length in zooplankton food webs. Ecology, 69(2): 418-426

Sugihara G, Schoenly K, Trombla A. 1989. Scale invariance in food web properties. Science, 245: 48-52

Valladares GR, Salvo A.1999. Insect-plant food webs could provide new clued for peat management. Environment Entomology, 28(4): 539-544

Wang KX, Zhang QQ, Chen LL, et al. 2013.Comparative studies on arthropod community structure characteristics between ecological rice paddies and conventional rice paddies. Plant Protection, 39(3): 31-35

Williams RJ andMartinez ND. 2000. Simple rules yieldcomplex food webs. Nature, 404:180-183

Wu JC, Hu GW, Tang J, et al. 1994. Studies on the regulation effect of neutral insect on the community food web in paddy field. Acta Ecologica Sinica, 14(4): 381-386

You MS, Pang XF. 1993. Trophic relations of arthropod community in paddy fields. Chinese Journal of Applied Ecology, 4(3): 278-282

Yuan W, Liu H, et al. 2010. Evaluation of communities of insect pests and natural enemies in organic rice fields of Changjiang Farm. Acta Agriculturae Shanghai, 26(2): 132-136

Zhang GR, Gu DX, Zhang WQ. 1996. Enzy-linked immunosorbent assay (ELISA) and its application in predation studies. Chinese Journal of Biological Control, 12(1): 33-38

Zhang RJ, He XF, Heong KL. 1997. Assessing impacts of climate change on paddy borer, *Scirpophaga incertula* (Walker) in Guangdong Province, China. Proceedings of the International Symposium on Integrated Pest Management in Rice - Based Ecosystem. 245-253, Guangzhou, China

Zhang WJ. 2007. Computer inference of network of ecological interactions from sampling data. Environmental Monitoring and Assessment, 124: 253-261

Zhang WJ. 2011. Constructing ecological interaction networks by correlation analysis: hints from community sampling. Network Biology, 1(2): 81-98

Zhang WJ. 2012a. Computational Ecology: Graphs, Networks and Agent-based Modeling. World Scientific, Singapore

Zhang WJ. 2012b. How to construct the statistic network? An association network of herbaceous plants constructed from field sampling. Network Biology, 2(2): 57-68

Zhang WJ. 2012c. Modeling community succession and assembly: A novel method for network evolution. Network Biology, 2(2): 69-78

Zhang WJ. 2012d. Several mathematical methods for identifying crucial nodes in networks. Network Biology, 2(4): 121-126

17

Decentralized control of ecological and biological networks through Evolutionary Network Control

Alessandro Ferrarini

Department of Evolutionary and Functional Biology, University of Parma, Via G. Saragat 4, I-43100 Parma, Italy

E-mail: sgtpm@libero.it, alessandro.ferrarini@unipr.it, a.ferrarini1972@libero.it

Abstract

Evolutionary Network Control (ENC) has been recently introduced to allow the control of any kind of ecological and biological networks, with an arbitrary number of nodes and links, acting from inside and/or outside. To date, ENC has been applied using a centralized approach where an arbitrary number of network nodes and links could be tamed. This approach has shown to be effective in the control of ecological and biological networks. However a decentralized control, where only one node and the correspondent input/output links are controlled, could be more economic from a computational viewpoint, in particular when the network is very large (i.e. big data). In this view, ENC is upgraded here to realize the decentralized control of ecological and biological nets.

Keywords centralized control; decentralized control; dynamical networks; genetic algorithms; Evolutionary Network Control; edge control; node control; network optimization; system dynamics.

1 Introduction

Evolutionary Network Control (ENC) has been recently introduced to allow the control of any kind of ecological and biological networks, with an arbitrary number of nodes and links, from inside (Ferrarini, 2013) and from outside (Ferrarini, 2013b). The endogenous control requires that the network is optimized at the beginning of its dynamics so that it will inertially go to the desired state. The exogenous control requires that one or more exogenous controllers act upon the network at each time step.

ENC can be applied to both discrete-time (i.e., systems of difference equations) and continuous-time (i.e., systems of differential equations) networks. ENC opposes the common idea in the scientific literature that controllability of networks should be based on the identification of the set of driver nodes that can guide the system's dynamics, in other words on the choice of a subset of nodes that should be selected to be permanently controlled (Ferrarini, 2011).

ENC makes use of an integrated solution (system dynamics - genetic optimization - stochastic simulations) to compute uncertainty about network control (Ferrarini, 2013c) and to compute control success and feasibility (Ferrarini, 2013d). ENC employs intermediate control functions to locally (step-by-step) drive ecological and biological networks, so that also intermediate steps (not only the final state) are under its control (Ferrarini, 2014). ENC can also globally subdue nonlinear networks (Ferrarini, 2015), impose early or late stability to any kind of ecological and biological network (Ferrarini, 2015b) and locally control nonlinear networks (Ferrarini, 2016).

Table 1 Evolutionary Network Control (ENC) and its developed variants.

Reference	Goal
Ferrarini 2011	Theoretical bases of Evolutionary Network Control
Ferrarini 2013	Endogenous control of linear ecological and biological networks
Ferrarini 2013b	Exogenous control of linear ecological and biological networks
Ferrarini 2013c	Computing the uncertainty associated with network control
Ferrarini 2013d	Computing the degree of success and feasibility of network control
Ferrarini 2014	Local control of linear ecological and biological networks
Ferrarini 2015	Global control of nonlinear ecological and biological networks
Ferrarini 2015b	Imposing early/late stability to linear and nonlinear networks
Ferrarini 2016	Local control of nonlinear ecological and biological networks
Ferrarini 2016b	Multipurpose control of ecological and biological networks
This work	Decentralized control of ecological and biological networks

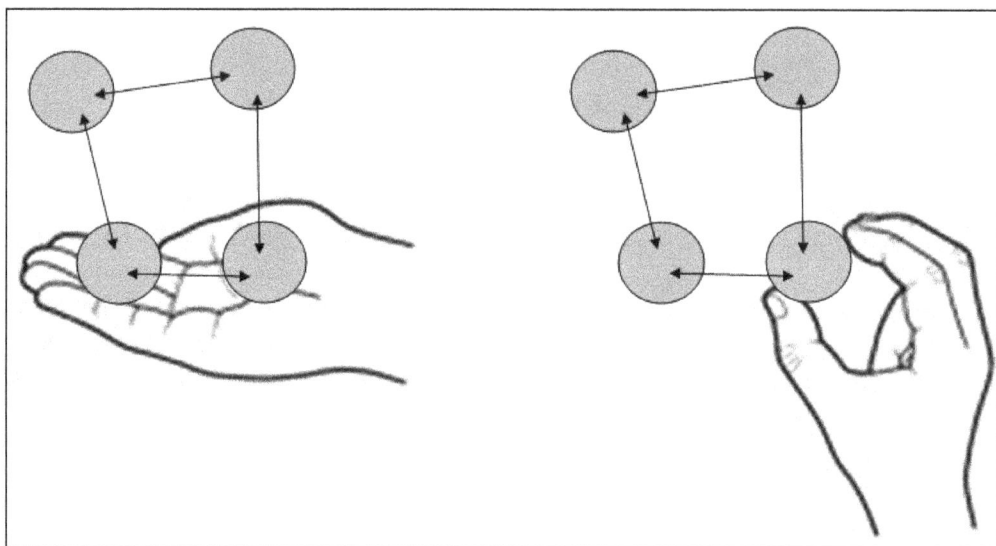

Fig. 1 Conceptual difference between centralized (on the left) and decentralized (on the right) control of ecological and biological networks. In the former case, the control can be applied to any node and link. In the latter case, only one node and the correspondent links are controlled.

To date, ENC has been applied using a centralized approach where an arbitrary number of network nodes and links could be tamed. This approach has shown to be effective in the control of ecological and biological networks. However a decentralized control, where only one node and the correspondent input/output links are controlled, could be more economic from a computational viewpoint, in particular when the network is very large, like when dealing with big data. In this view, ENC is upgraded here to realize the decentralized control of ecological and biological nets.

2 Decentralized Evolutionary Network Control: Mathematical Formulation

An ecological (or biological) dynamical system of n interacting taxonomic resolutions (species, genera, family, etc.) or aggregated assemblages of taxa (e.g., phytoplankton) is as follows

$$\frac{d\mathbf{S}}{dt} = \gamma(\mathbf{S}(t)) \tag{1}$$

where S_i is the number of individuals (or the total biomass) of the generic i-th taxonomic resolution (species, genera, family, or aggregated assemblages of taxa). If we also consider inputs (e.g. species reintroductions) and outputs (e.g. hunting) from outside, we must write:

$$\frac{d\mathbf{S}}{dt} = \gamma(\mathbf{S}(t)) + \mathbf{I}(t) + \mathbf{O}(t) \tag{2}$$

As noted by numerous authors (Luenberger 1979; Slotine and Li 1991) most real systems' dynamics can be modelled and simulated using a system of canonical, linear equations which represents a simplification of Eq. (2) as follows

$$\begin{cases} \frac{dS_1}{dt} = \alpha_{11}S_1 + ... + \alpha_{1n}S_n + I_1 + O_1 \\ ... \\ \frac{dS_n}{dt} = \alpha_{n1}S_1 + ... + \alpha_{nn}S_n + I_n + O_n \end{cases} \tag{3}$$

that can be written in the compact form

$$\frac{dS_i}{dt} = \underbrace{\alpha_{ii}S_i}_{\text{intra-specific}} + \underbrace{\sum_{j \neq i}\alpha_{ij}S_j}_{\text{inter-specific}} + \underbrace{\sum_k I_k + \sum_k O_k}_{\text{exogenous input-output}} \tag{4}$$

with initial values

$$\mathbf{S}_0 = <S_1(0), S_2(0)...S_n(0)> \tag{5}$$

and co-domain limits

$$\begin{cases} S_{1min} \leq S_1(t) \leq S_{1max} \\ ... \\ S_{nmin} \leq S_n(t) \leq S_{nmax} \end{cases} \tag{6}$$

and where

$$A = \begin{pmatrix} a_{11} & \dots & a_{1n} \\ \dots & \dots & \dots \\ a_{n1} & \dots & a_{nn} \end{pmatrix} \tag{7}$$

is the matrix of the per unit time effect on S_i due to unitary S_j. This system is inherently accelerated with acceleration equal to

$$\frac{d^2 S_i}{d^2 t} = \alpha_{ii} \frac{dS_i}{dt} + \sum_{j \neq i} \alpha_{ij} \frac{dS_j}{dt} + \sum_k \frac{dI_k}{dt} + \sum_k \frac{dO_k}{dt} \tag{8}$$

ENC solves the control of Eq. 3 using the following centralized approach (Ferrarini, 2013)

$$\begin{cases} \frac{dS_1}{dt} = a_{11*} S_1^* + \dots + a_{1n*} S_n^* + I_{1*} + O_{1*} \\ \dots \\ \frac{dS_n}{dt} = a_{n1*} S_1^* + \dots + a_{nn*} S_n^* + I_{n*} + O_{n*} \end{cases} \tag{9}$$

where any component (variable, parameter or coefficient) of Eq. 9 can be tamed, as denoted by the asterisk, to drive the network to the desired state. ENC makes use of genetic algorithms (GA; Holland, 1975; Goldberg, 1989) which consist of optimization procedures based on principles inspired by natural selection. GA involves 'chromosomal' representations of proposed problem solutions which undergo genetic operations such as selection, crossover and mutation.

ENC can also use an exogenous centralized control using an external controller C_1 (Ferrarini, 2013b)

$$\begin{cases} \frac{dS_1}{dt} = a_{11} S_1 + \dots + a_{1n} S_n + I_1 + O_1 + c_{11*} C_{1*} \\ \dots \\ \frac{dS_n}{dt} = a_{n1} S_1 + \dots + a_{nn} S_n + I_n + O_n + c_{n1*} C_{1*} \\ \frac{dC_1}{dt} = f_1 S_1 + \dots + f_n S_n \end{cases} \tag{10}$$

where asterisks stand for the genetic optimization of exogenous node's edges (i.e., coefficients of interaction with the inner system) and exogenous node's stock, i.e. the modification of such values at the beginning of network dynamics in order to get a certain goal (e.g., maximization of the final value of a certain variable). The controller C_1 that can also receive feedbacks from the network. that could be subject to control by taming $<f_1 \dots f_n>$.

In case 1 controller is not enough, the model in (10) must be expanded to the following k-external-controllers model (Ferrarini, 2013b):

$$\begin{cases} \dfrac{dS_1}{dt} = a_{11}S_1 + \ldots + a_{1n}S_n + I_1 + O_1 + c_{11*}C_{1*} + \ldots + c_{1k*}C_{k*} \\[2ex] \ldots \\[1ex] \dfrac{dS_n}{dt} = a_{n1}S_1 + \ldots + a_{nn}S_n + I_n + O_n + c_{n1*}C_{1*} + \ldots + c_{nk*}C_{k*} \\[2ex] \dfrac{dC_1}{dt} = f_{11}S_1 + \ldots + f_{1n}S_n \\[2ex] \ldots \\[1ex] \dfrac{dC_k}{dt} = f_{k1}S_1 + \ldots + f_{kn}S_n \end{cases} \qquad (11)$$

However, many ecological (or biological) dynamical systems can be more properly described using difference (recurrent) equations rather than differential ones. This is true for many systems where dynamics happen on discrete time rather than on continuous one.

In this case, Eq. 3 becomes

$$\begin{cases} \left(S_1\right)_{t+1} = a_{11}\left(S_1\right)_t + \ldots + a_{1n}\left(S_n\right)_t + \left(I_1\right)_t + \left(O_1\right)_t \\[1ex] \ldots \\[1ex] \left(S_n\right)_{t+1} = a_{n1}\left(S_1\right)_t + \ldots + a_{nn}\left(S_n\right)_t + \left(I_n\right)_t + \left(O_n\right)_t \end{cases} \qquad (12)$$

ENC solves the control of Eq. 12 using the following centralized approach

$$\begin{cases} \left(S_1\right)_{t+1} = a_{11*}\left(S_1\right)_t + \ldots + a_{1n*}\left(S_n\right)_t + \left(I_1\right)_{t*} + \left(O_1\right)_{t*} \\[1ex] \ldots \\[1ex] \left(S_n\right)_{t+1} = a_{n1*}\left(S_1\right)_t + \ldots + a_{nn*}\left(S_n\right)_t + \left(I_n\right)_{t*} + \left(O_n\right)_{t*} \end{cases} \qquad (13)$$

where asterisks stand for the optimization of edges (i.e., coefficients of interaction among variables) or nodes (i.e., initial stocks), that is the modification of their values at the beginning of the network dynamics in order to get a certain goal (e.g., maximization of the final value of a certain variable).

In this centralized approach (i.e. Eqs. 9, 10, 11 and 13) an arbitrary number of network nodes and links can be tamed.

Instead, in decentralized ENC only one node and the correspondent input/output links are controlled, this being true for all the previous driver equations (Eqs. 9, 10, 11 and 13). In this sense, decentralized ENC can be considered a particular case, computationally very reasonable, of the more general approach used by centralized ENC.

3 An Applicative Example

Fig. 2 depicts an *in silico* simulation of a real ecological network. The goal is to preserve target species' occurrence (centre of the network) in the study area.

Greenish nodes represent positive actors or events for the increase or preservation of the target species. Reddish nodes represent ecological actors or events with negative impact on the target species. Bluish nodes represent resources needed by the target species. Stocks stand for the actual amounts of individuals or biomass. Updates stand for yearly internal dynamics (i.e., intra-specific gains due to births and/or immigration rates minus losses due to deaths and/or emigration rates). Minimum and maximum values stand for lowest and highest values of stock values. For the sake of simplicity, the maximum possible value for each actor has been set to 100. Hunters and poachers remain constant (i.e. 10) during the simulation.

The percent value associated to links represent the percentage of the receiver that is yearly consumed by the transmitter at the beginning of the network simulation. Traps mortality and re-introductions accounts for 19 and 10 individuals per year respectively. It should be noted that predators can also gain resources from outside, so their internal dynamics (updates) are not limited to the presence of the target species.

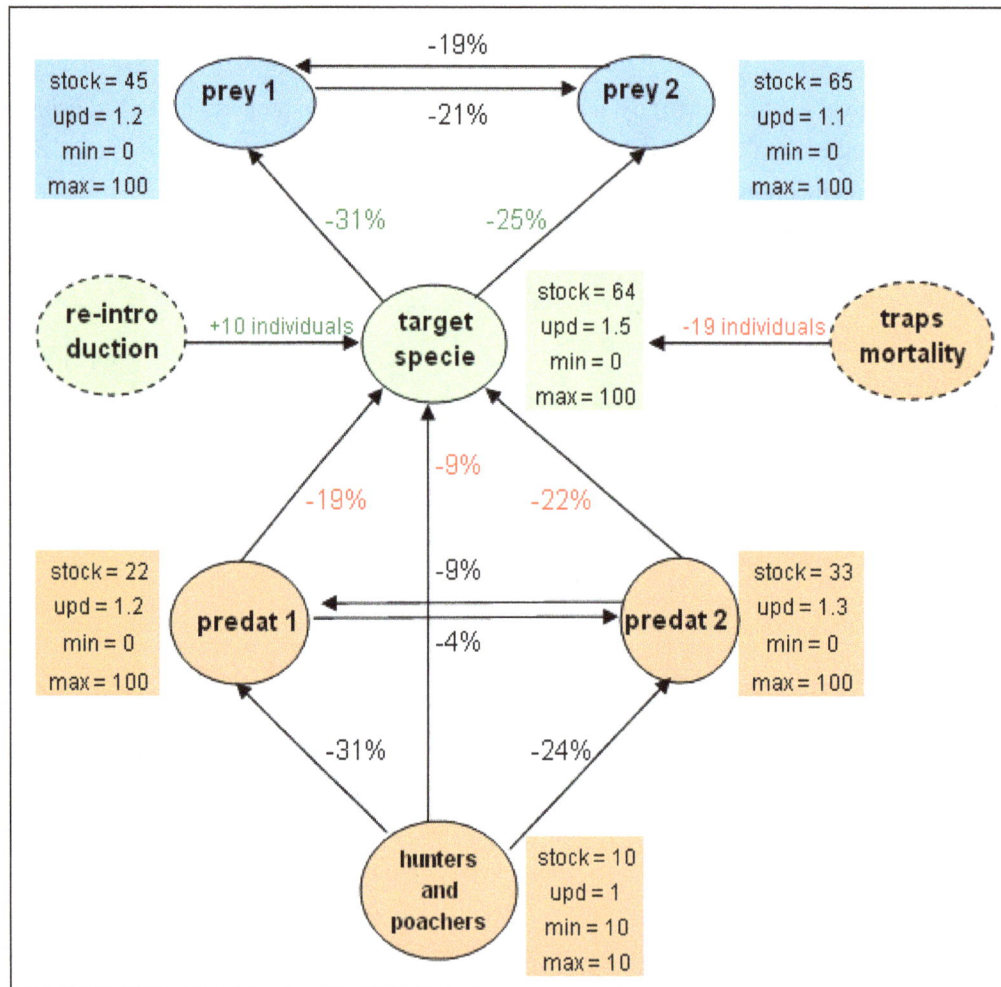

Fig. 2 Initial conditions of the ecological network under study.

The previous ecological network has the following inertial dynamics (Fig. 3), with equilibrium time $E_T =$ 22 years and the *target species* (green line) disappearing in the studied system after 5 years. The final vector at $E_T = 22$ is given by <*prey1*=100, *prey2*=0, *target*=0, *predat1*=0, *predat2*=100>.

Table 1 Parameter declaration corresponding to the ecological network of Fig. 2, calculated using Control-Lab 6 (Ferrarini, 2015c).

INITIAL STOCK VALUES						
	prey1	prey2	target	predat1	predat2	hunters
initial stock	45	65	64	22	33	10
min	0	0	0	0	0	10
max	100	100	100	100	100	10

INTERACTION MATRIX (yearly flows to receivers per unit of transmitters)							
		Receivers					
		prey1	prey2	target	predat1	predat2	hunters
Transmitters	prey1	1.2	-0.303	0	0	0	0
	prey2	-0.131	1.1	0	0	0	0
	target	-0.218	-0.254	1.5	0	0	0
	predat1	0	0	-0.552	1.2	-0.06	0
	predat2	0	0	-0.426	-0.06	1.3	0
	hunters	0	0	-0.576	-0.682	-0.792	1

yearly constant flows						
	Receivers					
prey1	prey2	target	predat1	predat2	hunters
0	0	-9	0	0	0

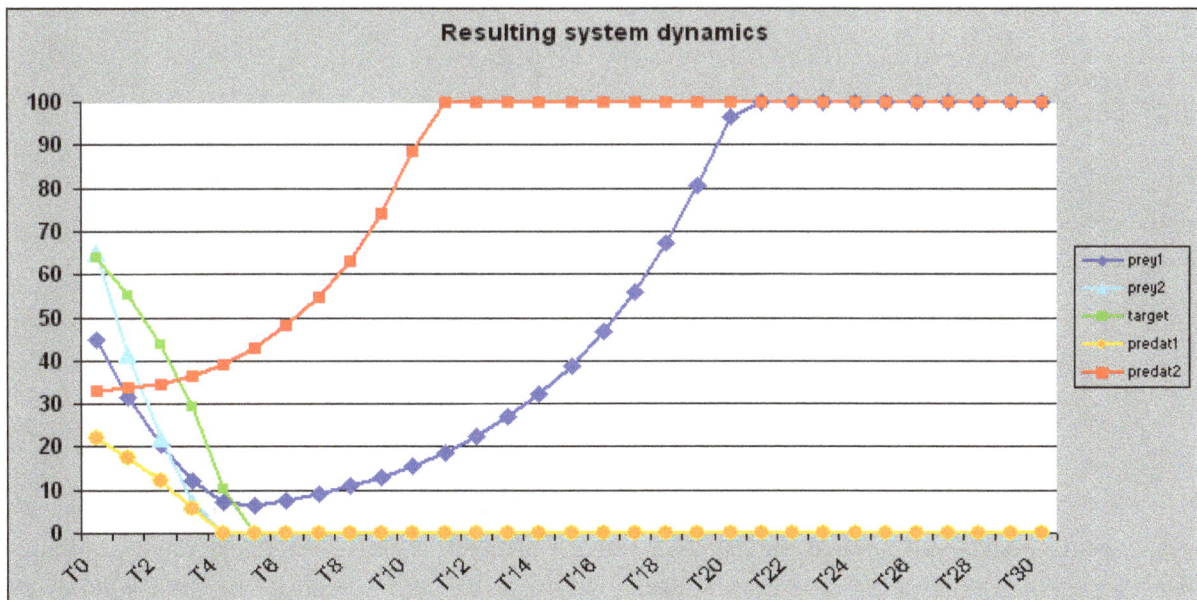

Fig. 3 Resulting dynamics for the network of Fig. 2. X-axis measures time in years. Dynamics have been calculated using the software Control-Lab 6 (Ferrarini, 2015c).

Now let's suppose we want to safeguard the target species by acting only upon *predat2*. I'll use a decentralized control where only the node *predat2* and its links will be tamed in order to have the target species at its maximum (i.e. 100) once that system dynamics reach the equilibrium. To make the control more realistic, it's opportune that the optimization of the node *predat2* and its links is limited to small changes (e.g. no more than 10% or no more than 20%).

Table 2 Ten solutions found by ENC to drive the system dynamics so that the target species reaches 100 at equilibrium. In this decentralized approach, ENC has only worked on node *predat2* and on its two links with the target species and with the other predator (*predat1*). The remaining network parameters were kept as in Tab. 1. Solutions have been detected using the software Control-Lab 6 (Ferrarini, 2015c).

detected solution	stock value of predator2	self coefficient of predator 2	interaction upon predator1	interaction upon target species	equilibrium value of target species	control achieved?
1	31	1.22	-0.07	-0.36	100	yes
2	33	1.09	-0.05	-0.40	100	yes
3	32	1.12	-0.04	-0.33	100	yes
4	32	1.10	-0.05	-0.30	100	yes
5	30	1.16	-0.08	-0.42	100	yes
6	31	1.10	-0.05	-0.36	100	yes
7	32	1.17	-0.08	-0.39	100	yes
8	30	1.01	-0.07	-0.31	100	yes
9	30	1.15	-0.05	-0.33	100	yes
10	30	1.26	-0.05	-0.34	100	yes

ENC has found many possible solutions to the decentralized control of the ecological network presented in Fig. 2, ten of which are described in Table 2. All these solutions lead to the same result, i.e. the target species reaches 100 at equilibrium.

It is clear that the decentralized ENC can be applied to any other actor of the network. In addition, one could seek a decentralized solution so that the ecological network reaches the desired solution within a predetermined time interval. For instance, by acting only upon *predat2* we could seek the control of the ecological network of Fig. 2 so that the target species reaches 100 with equilibrium at $T<10$ years. One possible solution detected by decentralized ENC is depicted in Fig. 4.

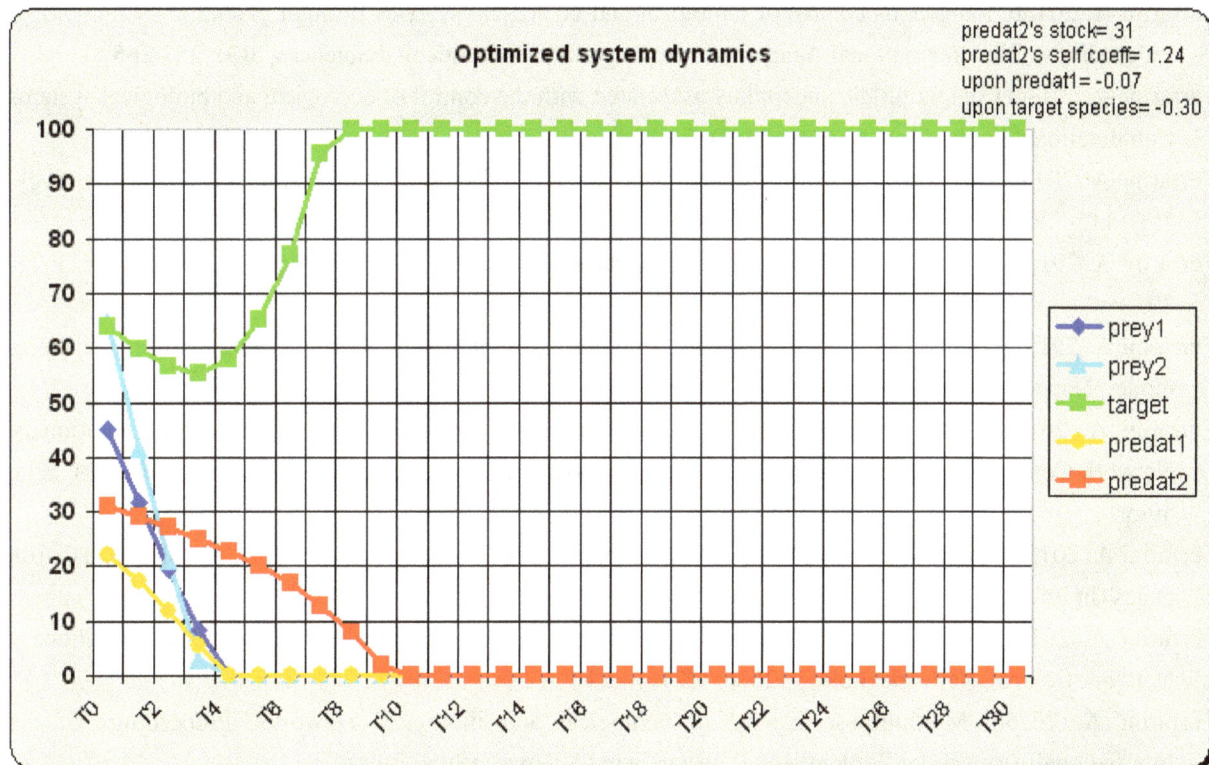

Fig. 4 A solution detected by ENC to drive the system dynamics so that the target species reaches 100 with equilibrium at $T<10$. In this decentralized approach, ENC has only worked on node *predat2* and on its two links with the target species and with the other predator (*predat1*). The detected model parameters are described on the top right. Solutions have been detected using the software Control-Lab 6 (Ferrarini, 2015c).

The framework proposed here might also be applied to semi-quantitative networks (Ferrarini, 2011b). Decentralized ENC has been applied using the software Control-Lab 6 (Ferrarini, 2015c) written in Visual Basic (Balena, 2001; Pattison, 1998).

4 Conclusion

Evolutionary network control (ENC) has been introduced as a centralized methodology where an arbitrary number of network nodes and links could be tamed to drive the network dynamics towards the desired outputs. ENC has shown to be very effective in the control of ecological and biological networks. However a decentralized control, where only one node and the correspondent input/output links are controlled, could be more economic from a computational viewpoint, in particular when the network is very large. In this sense, decentralized ENC results very promising when applied to big data, the new frontier of network dynamics and control.

References

Balena F. 2001. Programming Microsoft Visual Basic 6.0. Microsoft Press, Redmond, WA, USA
Ferrarini A. 2011. Some thoughts on the controllability of network systems. Network Biology, 1(3-4): 186-188
Ferrarini A. 2011b. Some steps forward in semi-quantitative network modelling. Network Biology, 1(1): 72-78
Ferrarini A. 2013. Controlling ecological and biological networks via evolutionary modelling. Network Biology, 3(3): 97-105

Ferrarini A. 2013b. Exogenous control of biological and ecological systems through evolutionary modelling. Proceedings of the International Academy of Ecology and Environmental Sciences, 3(3): 257-265

Ferrarini A. 2013c. Computing the uncertainty associated with the control of ecological and biological systems. Computational Ecology and Software, 3(3): 74-80

Ferrarini A. 2013d. Networks control: introducing the degree of success and feasibility. Network Biology, 3(4): 115-120

Ferrarini A. 2014. Local and global control of ecological and biological networks. Network Biology, 4(1): 21-30

Ferrarini A. 2015. Evolutionary network control also holds for nonlinear networks: Ruling the Lotka-Volterra model. Network Biology, 5(1): 34-42

Ferrarini A. 2015b. Imposing early stability to ecological and biological networks through Evolutionary Network Control. Proceedings of the International Academy of Ecology and Environmental Sciences, 5(1): 49-56

Ferrarini A. 2015c. Control-Lab 6: a software for the application of Ecological Network Control. Manual, 108 pages (in Italian)

Ferrarini A. 2016. Bit by bit control of nonlinear ecological and biological networks using Evolutionary Network Control. Network Biology, 2016, 6(2): 47-54

Ferrarini A. 2016b. Multipurpose control of ecological and biological networks. Proceedings of the International Academy of Ecology and Environmental Sciences (Submitted)

Goldberg DE. 1989. Genetic Algorithms in Search Optimization and Machine Learning. Addison-Wesley, Reading, USA.

Holland JH. 1975. Adaptation in Natural and Artificial Systems: An Introductory Analysis with Applications to Biology, Control and Artificial Intelligence. University of Michigan Press, Ann Arbor, USA

Luenberger DG. 1979. Introduction to Dynamic Systems: Theory, Models, & Applications. Wiley, USA

Pattison T. 1998. Programming Distributed Applications with COM and Microsoft Visual Basic 6.0. Microsoft Press, Redmond, WA, USA

Slotine JJ, Li W. 1991. Applied Nonlinear Control. Prentice-Hall, USA

18

Protein and mRNA levels support the notion that a genetic regulatory circuit controls growth phases in *E. coli* populations

Agustino Martínez-Antonio

Departamento de Ingeniería Genética. Centro de Investigación y de Estudios Avanzados del Instituto Politécnico Nacional. Unidad Irapuato. Km. 9.6 Libramiento Norte Carretera Irapuato-León. CP 36821. Irapuato, Guanajuato. México.
E-mail: amartinez@ira.cinvestav.mx

Abstract

Bacterial populations transition between growing and non-growing phases, based on nutrient availability and stresses conditions. The hallmark of a growing state is anabolism, including DNA replication and cell division. In contrast, bacteria in a growth-arrested state acquire a resistant physiology and diminished metabolism. However, there is little knowledge on how this transition occurs at the molecular level. Here, we provide new evidence that a multi-element genetic regulatory circuit might work to maintain genetic control among growth-phase transitions in *Escherichia coli*. This work contributes to the discovering of design principles behind the performance of biological functions, which could be of relevance on the new disciplines of biological engineering and synthetic biology.

Keywords growth phase; regulatory circuit; proteome; mRNA; bacteria.

1 Introduction

It is well known that bacteria multiply rapidly when nutrients are plentiful and arrest their growth when carbon sources are depleted or stresses conditions occur. When bacteria are grown in batch culture, the population follows a well-defined curve with previously described growth phases (Monod, 1949). These growth phases have been modeled mathematically (Zwietering et al., 1990). One can assume that bacteria arrest their growth when nutrients are limiting and resume their growth when conditions are favorable again (Kolter et al., 1993). This simple supposition implies that the cellular machinery is designed to function in a continuous mode; designed to arrest and re-initiate function depending upon nutrient availability and/or stress conditions. Nevertheless, biochemical and genetics studies provide us with clues about the molecular processes that occur when bacteria transition between active and arrested growth. However, the molecular details of this mechanism are more complex than this simple conjecture implies. In fact, bacteria need to adapt their cellular machinery to changing conditions; this adaptation includes the altering of transcriptional expression profile.

The phenotypic result of these molecular changes is transition of a bacterial population between growth phases.

Methods in molecular biology have enabled us to identify hundreds of genes and, in some cases, the regulators that control their expression. The most precise experiments that link the activity of regulators to their target genes are those that are investigated specifically and individually. These types of studies produce detailed results on the regulatory interactions between one regulator and one target gene. This information is gathered and curated on dedicated databases, such as RegulonDB (Salgado et al., 2013). The activation and repression of gene transcription is a task executed by the regulatory machinery, which includes nucleoid-associated proteins, sigma factors, and transcription factors that operate in an intricate regulatory network (Martinez-Antonio, 2011).

Previously, we described a multi-element genetic regulatory circuit that may be implicated in controlling the transition between growths phases in *Escherichia coli* (Martinez-Antonio et al., 2012). In that study, we described the components of the genetic regulatory circuit and offered a rationale for this hypothesis. Additionally, we developed a mathematical model, consisting of differential equations based on power-law formalism, to determine how this circuit might be operating. Here, we searched transcription and proteome data that could lend further support to this hypothesis. We show that mRNAs and proteins corresponding to the regulators on the network are more abundantly expressed at times that corresponds to their peak of activity within the growth phases circuit.

2 Materials and Methods

2.1 Regulatory interaction data

The pairwise transcriptional regulatory interactions between genes and regulators were obtained from RegulonDB v8.0 (Salgado et al., 2013). To reduce the network, nodes corresponding to non-regulatory genes were eliminated; however, the primary network of regulatory genes was kept intact. From this last subset of nodes and interactions, the regulators forming the circuit were extracted, as shown in Fig. 1.

2.2 Transcriptome data

Data on the mRNA levels of genes on the circuit were searched at the NCBI GEO database (Barrett et al., 2011). Care was taken to ensure that included information was not generated by experiments using gene deletion, gene over-expression, environmental stressors, or any other condition that could mask or influence the presence of transcripts beyond that of the normal transition of bacteria between growth phases. Useful data corresponding to the genes in the circuit were extracted manually.

2.3 Proteome data

Due the scarcity of this type of data on dedicated databases, proteome data were mined from the original literature on PubMed (http://www.ncbi.nlm.nih.gov/pubmed/). The key words "proteome data" and "*E. coli*" were used for these searches. Using the same inclusion criteria as above for mRNA data, useable data corresponding to the proteins in the circuit were extracted manually from primary and supplementary figures within the articles.

3 Results

3.1 The genetic regulatory circuit controlling growth phase in *E. coli*

When all experimentally validated, pairwise, regulatory interactions are combined, a number of multi-element regulatory circuits begin to emerge (Martinez-Antonio et al., 2008). One of these circuits (Fig. 1) involves global regulators at the core of the entire genetic regulatory network in *E. coli*. One multi-element genetic regulatory circuit, comprised a set of genes and regulators that activate and repress expression in a way that form a closed path, as shown in Fig. 1. A description of the components in the circuit is given in Table 1. Overall, this circuit is a negative feedback loop, designated by a negative sign; the products of the signs of its edges. This result means that the circuit displays homeostatic control and a periodic behavior.

Embedded within this genetic regulatory circuit are two additional regulatory circuits, one negative (HNS-GadX) and one positive (GadX-RpoS). Both embedded circuits have to GadX as the common element. GadX has been proposed as the master switch for the activity of this circuit because inactive GadX protein maintains activity of the HNS-GadX circuit, while active GadX shifts the activity of the circuit to RpoS and IHF (Martinez-Antonio et al., 2012). Dynamic studies on gene regulatory circuits reveal that circuits like the one described here could have multiple functions and complex behaviors if positive and negative circuits are embedded within them (Thomas et al., 1995). In other words, this kind of circuit can produce different steady states of gene expression patterns under different physiological conditions (Kaufman et al., 2007). In the case of this circuit, the biological implication of such a regulator switch is that the activity of these regulatory components and their functions may be linked to the various growth phases of this bacterium. In subsequent sections, we provide some evidence that this circuit is regulating gene expression in a growth phase-dependent manner in *E. coli*.

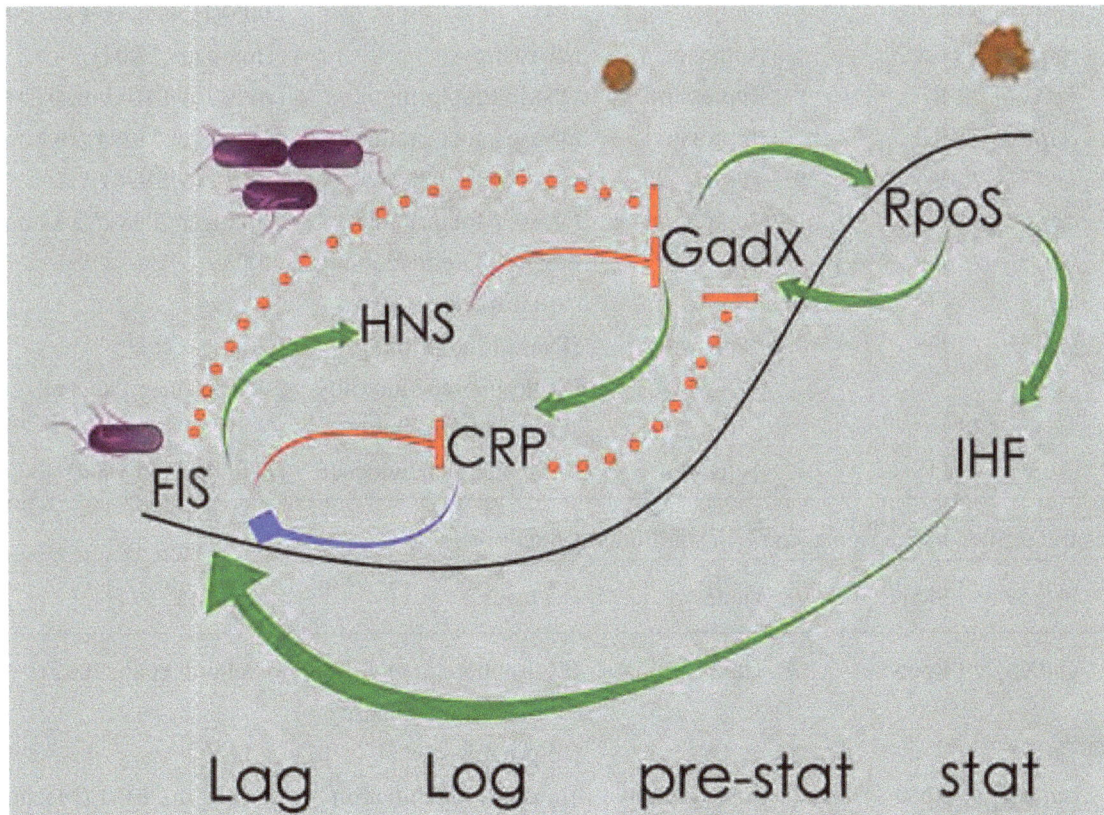

Fig. 1 The regulatory circuit controlling growth-phases in *E. coli*. This cartoon represents the growth phases and the regulators of the circuit, illustrated which is more active in each case: Green = represent activation; red = repression; blue = dual regulation (activation and repression). The grey line represents the bacterial growth curve.

Table 1 Validated regulatory interactions between elements of the circuit.

Gene	Promoter	Transcription Factor	Mode of regulation	Evidence	Reference
fis	*dusBp*	CRP-cAMP	Activation	Microarrays	(Zheng et al., 2004)
fis	*dusBp*	CRP-cAMP	Dual	DNase I footprinting	(Nasser et al., 2001)
fis	*dusBp*	FIS	Repression	DNase I footprinting	(Ball et al., 1992) (Hengenet al., 1997)
fis	*dusBp*	IHF	Activation	Site mutation, reporter assays	(Nasser et al., 2002)(Pratt et al., 1997)
hns	*hnsp*	FIS	Activation	DNase I footprinting	(Falconi et al., 1996)(Giangrossi et al., 2001)
hns	*hnsp*	Gadx	Activation	Microarrays, TF overexpression	(Hommais et al., 2004)
hns	*hnsp*	HNS	Repression	DNase I footprinting	(Falconi et al., 1996)(Falconi et al., 1993)(Giangrossi et al., 2001)(Ueguchi et al., 1993)
gadX	gadXp	GadX	Activation	Microarrays, RT-PCR	(Ma et al., 2002)(Hommais, 2004)(Tramonti et al., 2008)
gadX	gadXp	HNS	Repression	DNase I footprinting	(Giangrossi et al., 2005)(Hommais et al., 2001)
rpoS	rpoSp	GadX	Activator	Microarrays	(Hommais, 2004)
ihfA	*ihfAp4*	IHF	Repression	DNase I footprinting	(Aviv 1994)(Bykowski 1998)
ihfB	*ihfBp*	IHF	Repression	DNase I footprinting	(Aviv et al., 1994)(Bykowski and Sirko, 1998)
dps	*dpsp*	FIS	Repression	DNase I footprinting, Electrophoretic mobility shift, reporter assays	(Grainger 2008)(Yamamoto 2011)
dps	*dpsp*	FIS	Repression	DNase I footprinting, Electrophoretic mobility shift, reporter assays	(Grainger et al., 2008)(Yamamoto et al., 2011)
dps	*dpsp*	FIS	Activation	Computational evidence	(Altuviaet al., 1994)

Genes transcribed by the sigma RpoS, in addition to RpoD

Gene	Promoter	Sigma	Mode	Evidence	Reference
gadX	gadXp	RpoS	Transcription	Electrophoretic mobility shift assay and DNase I footprinting	(Tramonti et al., 2002)
ihfA	*ihfAp4*	RpoS	Transcription	Transcription initiation mapping	(Aviv et al., 1994)(Mechulam et al., 1987)
ihfB	*ihfBp*	RpoS	Transcription	Transcription initiation mapping	(Tramonti et al., 2002)(Węgleńska et al., 1996)

3.2 The regulatory factors of the circuit

Three of regulators in the circuit on Fig. 1 are nucleoid-associated proteins or NAPs (FIS, HNS and IHF) (Dillon, 2010). These proteins bend and bridge the DNA in different conformations. The Ishihama laboratory studied the abundance of NAPs in *E. coli*, primarily by western blot analysis. They reported that the NAPs present in this circuit were maximally expressed in a growth phases-dependent manner. First, FIS expression peaks when cells start to divide before the exponential phase. Next, HNS expression is maximal during exponential growth. Finally, IHF is expressed mostly in stationary phase (Azam et al., 1999). The circuit also contains the acid-stress resistance regulator, GadX. GadX belongs to a group of transcriptional regulators that respond to low pH, mainly due to intracellular acidification by the accumulation of organic acids resulting from fermentative metabolism (Tramontiet al., 2002; Ma et al., 2002). The circuit is completed with a general stress response sigma factor, RpoS, which replaces the activity of the housekeeping sigma factor RpoD during stress conditions. Transcription of the anti-sigma factor RSD, inactivates RpoD. RpoS is the master sigma factor that directs RNAP to the transcription of genes, including the promoters of IHF subunits, whose products respond to multiple stress types (Lange and Hengge-Aronis, 1991).

How might this circuit operates?

The main properties of regulators in the circuit and the functional classes of their regulons are shown in Table 2 and Fig. 2, with brief descriptions of each elements of the circuit (an additional, detailed description was presented in Martinez-Antonio et al. (2012)). FIS should be a very important player at the beginning of bacterial growth because it activates the transcription of important cellular elements dealing with the process of cell division, such as tRNAs, rRNAs, and stable RNAs, as well as ribosomal RNAs and genes for translation (Finkel and Johnson, 1992). Some of these same genes are also regulated by HNS (Free and Dorman, 1995), which is the regulator that follows FIS in the circuit. Interestingly, *hns* is activated by FIS. GadX regulates primarily the genes for pH homeostasis; most of these genes are co-regulated by HNS. *gadX* activates *hns* and is repressed by *hns*. This mutual regulation with opposed signs constitute a negative circuit of regulation. During ideal growth conditions, the inactive form of GadX should stop the main activity of the whole circuit at this point in the pathway.

GadX can be allosterically activated by organic acids; such as acetate and formate (Shin et al., 2001). Usually, the presence of such acids is indicative of acidic stress conditions, such as those produced by cells entering into fermentative metabolism. Organic acids activate GadX, which increases the transcription of *rpoS*. Because the *gadX* gene has a promoter for RpoS, a robust positive circuit forms. RpoS transcribes many genes that prepare the cell to acquire a resistant physiology, including those that induce a smaller, rounded morphology, such as the regulator BolA (Aldea et al., 1989). RpoS also transcribes the two of the IHF subunits *(ihfA* and *ihfB)* and IHF activates the transcription of *dps*, which encodes a small protein in late stationary phase that forms crystals with DNA to protect it (Altuvia et al., 1994). IHF regulates many genes, but notable for this discussion are those for anaerobic respiration. In addition, IHF activates *fis*, and with this interaction, the circuit is closed. At the DNA origin of replication in the *E. coli*, there is a DNA-binding site for IHF, which suggests that this regulator may be involved in this process. IHF may function by bending the DNA and preventing or facilitating the access of the replication machinery to the origin of DNA replication (Goosen and Van de Putte, 1995).

The overall activity of this circuit was modeled (Martinez-Antonio et al., 2012) and revealed that GadX might serve as a checkpoint of the circuit by maintaining the negative circuit while inactive and activating the positive circuit in response to organic acids. It is proposed that this circuit should contribute to the robust population-wide decision to continue or arrest growth. By activating the second part of the circuit, starting from GadX and RpoS, the regulatory machinery ensures that bacteria change the pattern of gene expression upon growth arrest. One can expect that an analogous checkpoint should exist to facilitate the transition from arrested growth to an active growing state. At this point in the circuit architecture, no such analogous switch has been found. Certainly, this transition could not be explained solely by the activation of *fis* by IHF, however, this transition might also depend on the control of CRP over *fis*. CRP is the most global regulator in *E. coli* and its activity depends allosterically on the presence of cAMP (Harman, 2001). It means that CRP could sense the overall energetic status of the cell, including information on the carbon sources availability, and might have the capability to activate or repress *fis*, thus controlling the decision for growth is when conditions are suitable.

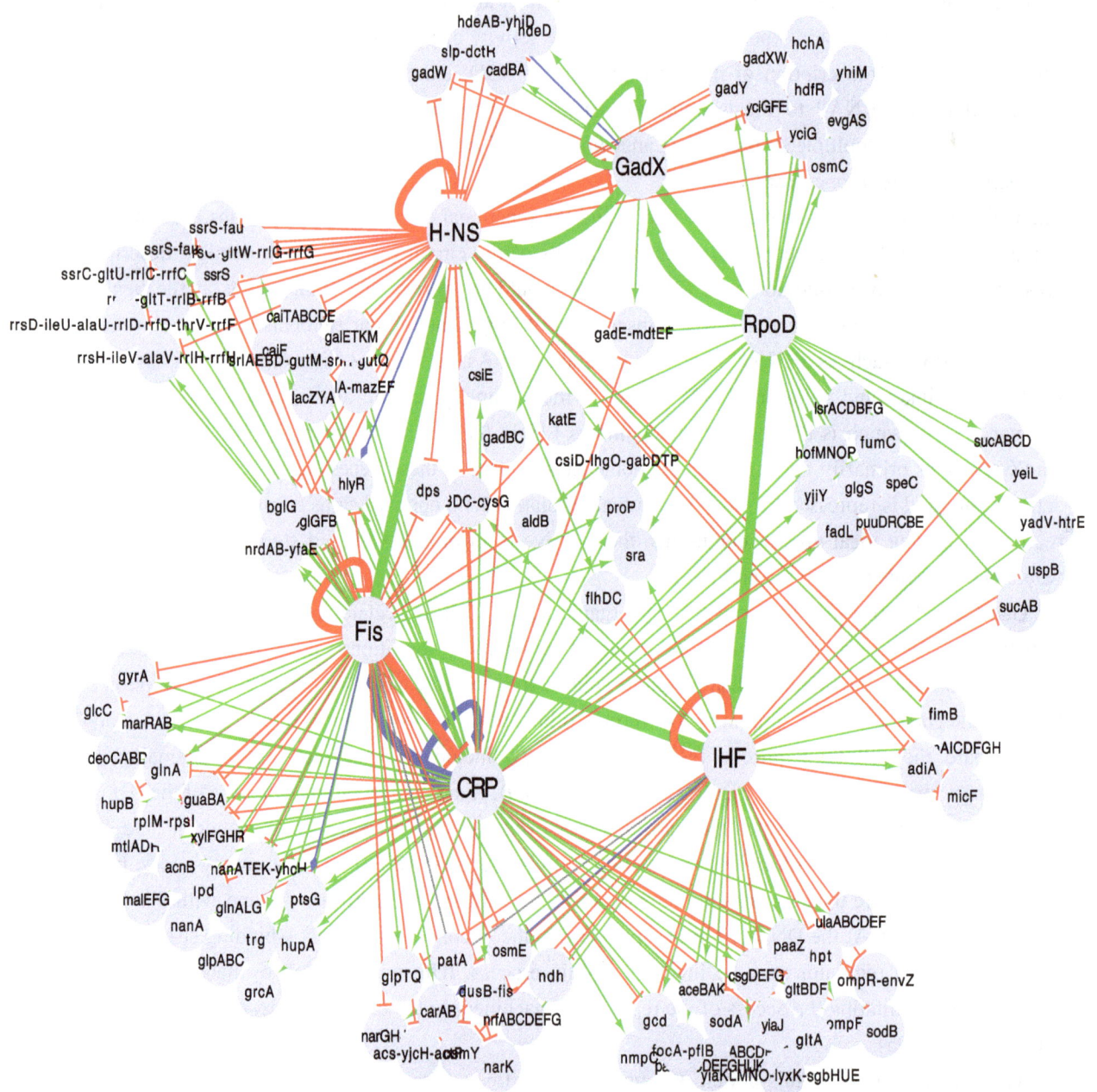

Fig. 2 Genes controlled by each regulator of the genetic circuit. Some of these regulated genes are also regulated by other transcription factors of *E. coli*, but for clarity, only the regulation exerted by regulators of this circuit are included here.

Table 2 The regulatory genes of the circuit and the functional classes of the regulated genes.

Transcription factor	Description	Functional classes of target genes (numbers), take from RegulonDB
FIS (Factor for Inversion Stimulation)	A 22 kDa homo-dimeric protein. FIS bends the DNA between 50° and 90°.	tRNAs (53), anaerobic respiration (34), membrane (34), translation (29), ribosome (27), aerobic respiration (23), rRNA and stable RNAs (22), carbon compounds (20), electron donors (20), transcription related (18).
HNS (Histone-like Nucleoid Structuring protein)	A 15.4 kDa protein that forms bridges between adjacent DNA duplexes.	Transcription related (24), carbon compounds (24), membrane (23), activators (20), translation (17), ribosomes (17), rRNA and stable RNAs (16), uncharacterized proteins (14), pH homeostasis (13).
GadX (Regulator of Glutamic Acid Decarboxylase)	Contributes to pH homeostasis by consuming intracellular H+ and producing gamma-amino butyric acid	pH homeostasis (8), Porters (5), membrane (5), transcription related (4), activators (3), amino acids (2).
RpoS (Sigma S or sigma38)	A sigma subunit of RNAP for general stresses and stationary phase transcription	Diverse stress-responses (60)
IHF (Integration Host Factor)	A protein composed of α (*himA*) and β (*himB*) subunits. It bends the DNA and compact the chromosome length by about 30%	Anaerobic respiration (42), membrane (41), carbon compounds (21), transcription related (19), aerobic respiration (16), electron donor (15), activators (14), porters (13), oxide-reduction transporters (13)

3.3 mRNA levels of the regulatory genes in the circuit support the circuit model

We searched the mRNA levels of genes on the circuit in the NCBI GEO database (Barrett et al., 2011). Ideally, the data used in this analysis should not be obtained from experiments that involve gene deletions, gene over-expression, environmental stress, or any other condition that could mask or influence the presence of transcripts beyond those that result from the natural transition of bacteria through growth phases. One such exceptionally useful study was published by Sangurdekar et al (2006). In this work, the authors measured mRNA abundance of a culture of *E. coli* MG1655 grown in the minimal medium Bonner-Vogel at 0.5 DO and compared the results to those obtained from the same strain grown in LB medium at multiple time points that covered all the growth phases. From this study, we could recover information about the mRNA abundance for five of the six regulatory genes of the regulatory circuit (since IHF is constituted by two genes: *ihfA* and *ihfB*). Absent information was for one of the subunits of the IHF protein (*ihfB*). This analysis revealed that the quantity of mRNA varies for each gene over the growth phases (Fig. 3). In the case of *fis* and *hns*, their transcripts are more abundant before mid-exponential phase. In contrast, the transcripts for *gadX*, *ihfA*, and *rpoS* are more abundant after mid-exponential phase. For comparison, we decided to look for the mRNA quantity of *dps* because it is expected to be abundant in the late stationary phase. We found that transcripts level of *dps* were most abundant in stationary phase, further supporting the hypothesis that this circuit modulates the transition into stationary phase.

To validate the accuracy of the data used for this analysis, we examined the expression pattern of several control genes with known expression changes over the entire growth curve. The repressor LacI is not required in these conditions (Semsey et al., 2013), and we observed no major changes in *lacI* mRNA. Topoisomerase 1 *(topA)* is supposed to be active during DNA replication to separate the DNA strands (Valjavec-Gratian et al., 2005). We noted more abundant mRNA of this gene from the beginning of growth through the first half of the exponential phase. Lastly, the global regulator CRP is subject to dual regulation, including self-regulation, was slightly more abundant in exponential phase (Fig. 3). Thus, at the mRNA level, these regulators in the circuit are differentially transcribed, likely because they are required at different bacterial growth phases.

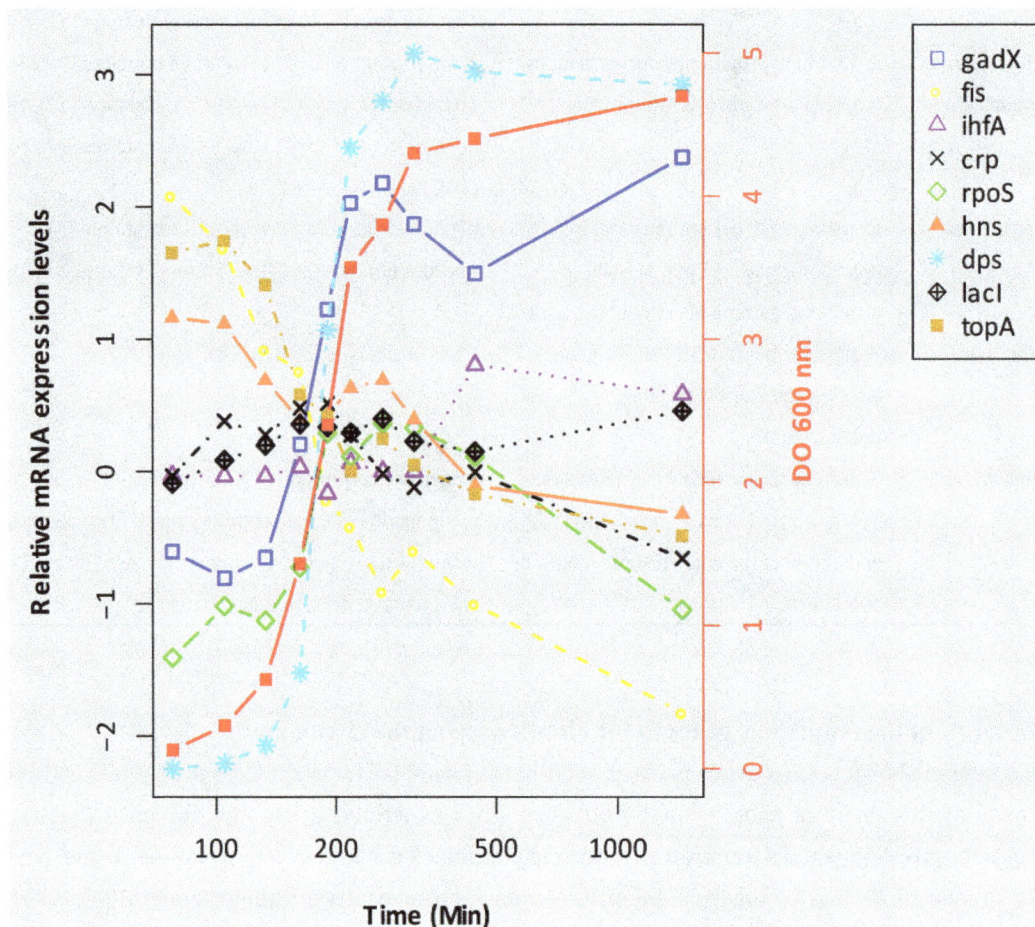

Fig. 3 mRNA profiles of transcription factors on the regulatory circuit. Relative quantities of mRNA are shown (left y-axis) over time (x-axis). The bacterial growth curve (red line, right Y-axis) is shown to illustrate growth phases. The strain used was K12 MG1655 grown in LB medium (Sangurdekar et al., 2006).

3.4 Protein levels of regulators in the circuit

Next, we examined the expression of regulators at the protein level. Despite exhaustive searching, we found only one proteomic study in minimal medium were the authors applied stable isotope labeling to amino acids in cell culture (SILAC) and performed a quantitative analysis of proteome dynamics in *E. coli* BW25113 during five distinct phases of growth (Soares et al., 2013). Our analysis is summarized in Fig 4. Soares et al. took as reference the quantity of proteins of a culture just entering the stationary phase ("point 4" on Fig. 4) and compared the relative abundance of proteins in the samples from other growth phases. With these data on hand, we looked for the relative protein abundance of the regulators in the circuit. We obtained information for four of the five regulators of the circuit (FIS, HNS, RpoS, IHF). Data for GadX were not available; thus, we used data from GadE, which is directly activated by GadX and is involved on the acid-stress response. Abundance of GadE may provide indirect information about the abundance of GadX. The profiles of relative protein abundance are shown in Fig. 4. FIS protein levels are more abundant when cells start to divide and enter into the exponential phase; and levels of FIS fall as the bacteria decelerate their rate of growth. The sigma factor RpoS and the two subunits of IHF (IHFA and IHFB) augment their quantities as the culture enters the stationary phase. The protein levels of HNS and CRP seem to be slightly more abundant at some points in the exponential phase; however, changes in the abundance of these proteins are less robust. Dps protein was more abundant in late stationary phase. Finally, the protein levels of TopA and LacI remain almost constant across all growth phases. Similar to the results seen with the mRNA analysis, the protein profiles of the elements of the circuit support the notion that these regulators should be more abundant when their activity peaks.

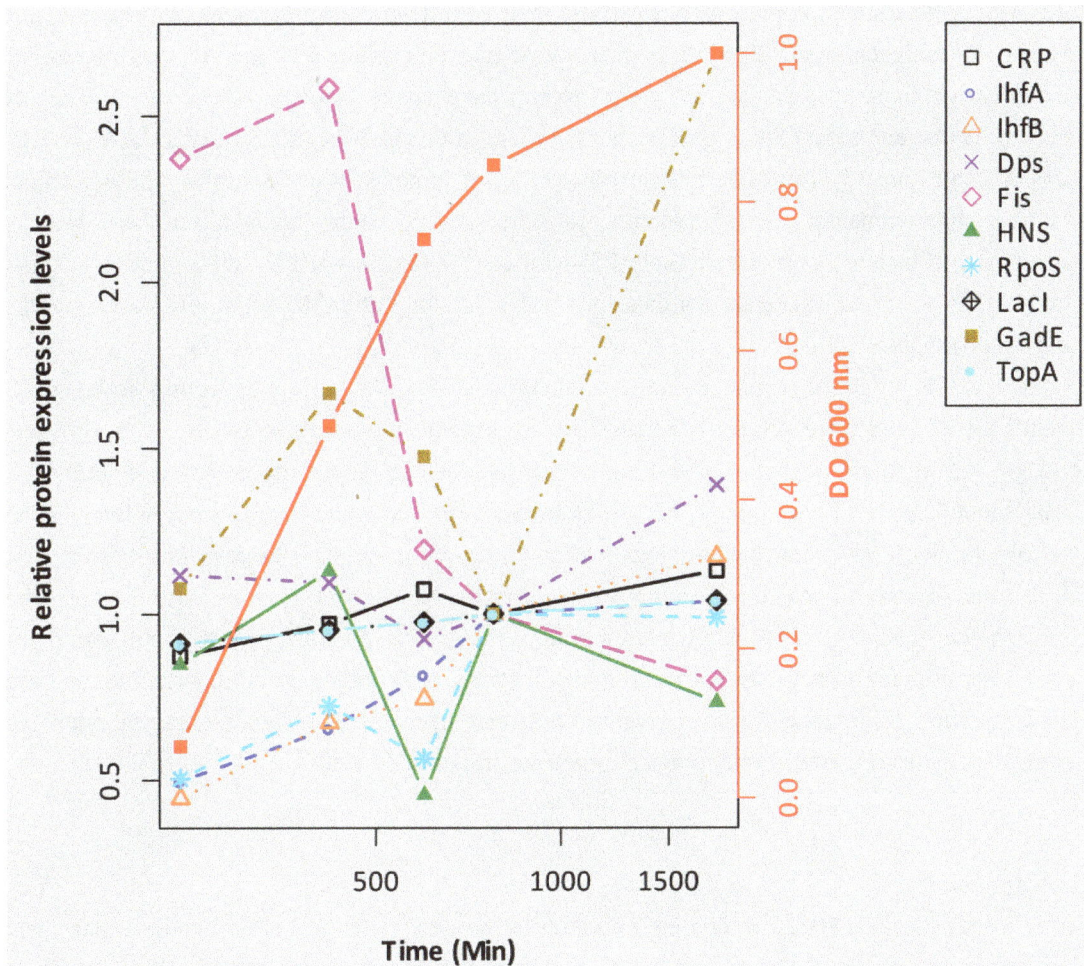

Fig. 4 Relative protein levels of regulators in the circuit. Relative quantities of proteins are shown (left y-axis) over time. The bacterial growth curve (red line, right y-axis) is shown to illustrate growth phases. The strain used was W3110 grown in M9 minimal medium (Soares et al., 2013).

4 Discussions

In this study, we describe a multi-element genetic regulatory circuit whose components function to provide control, fitness, and robustness to the process of population growth of *E. coli*. The elements and architecture of the circuit are organized such that they regulate each other in periodic fashion that makes biological sense. This circuit is arranged into two smaller parts, operating in either active growth or growth-arrested conditions. Given what is known about the architecture and biological roles of each regulator of the circuit, it makes biological sense they control population growth. Here, we provide proteomic and transcriptomic data that support this hypothesis of growth regulation. The presence and abundance of mRNA and proteins of these components peaked when they are more active.

It is rare to find proteomic and transcriptomic data at several phases of growth for the same culture; fortunately we found data that, although generated for other purposes, served nicely for this analysis. With these data, we confirmed the notion that the maximal abundance of these elements occurs when these regulators should be required, offering a form of temporal support for this hypothesis. The architecture and proposal activity of the genetic regulatory circuit could explain how it operates to start and arrest bacterial growth.

Although mRNA was reported in relative units; however, to appreciate their small quantity, studies on the total mRNA have determined a median value of less than 10 mRNA copies per gene per cell in a single-cell study on *E. coli* (Taniguchi et al., 2010). This observation may be explained by the fact that mRNA is quickly degraded, often within minutes. In contrast, many proteins have half-lives greater than the *E. coli* cell cycle. In the case of proteins, one recent study by (Wiśniewskia and Rakusb, 2014) stated that *E. coli* (ATCC 25922 grew at 37°C, 250 rpm, 15 hrs in LB medium) has 75 fg of proteins in late stationary phase. This number corresponds to approximately 1.3×10^6 proteins/cell. The specific values on late stationary phase for the proteins referred to here are in molecules/cell: 2534 for IHFA, 1582 for IHFB; 2980 for CRP; 206 for FIS; 6059 for HNS; 55 for GadE (there are not data for GadX); 1.7 for RpoS (101 for RpoD); 9 for LacI; 125 for TopA and 4339 for Dps.

Isogenic mutants for all the regulators in the circuit are available in the Keio collection (Baba et al., 2006). The mutants for FIS and RpoS are the most sensitive for growth in our hands; a strain with a *fis* deletion is unable to grow in minimal medium without additional supplements, including carbon and nitrogen sources (e.g., casamino acids). The RpoS mutant was lost from the collection with three successive freezings to -80°C (personal observation). With new methodologies at hand, now it is possible to determine the importance of intracellular macro-components to physiological requirement, such as ribosomes, mRNA, proteins, etc., depending on the quality of nutrients (Klumpp et al., 2009). It is possible that regulators of this circuit, although not essential for bacterial growth in normal conditions, are evolutionarily important to the fitness of *E. coli*. Our description of this kind of circuit reveals a potential mechanism to explain observed phenotypes and could guide the engineering of certain biological processes in synthetic biology.

Acknowledgment

The author thanks Mariana Heras for helping to search for transcriptome data and Edgardo Galán Vásquez for his help on the elaboration of Figures 2 and 3. This work was supported by the CONACYT grant 103686.

References

Aldea M, Garrido T, Hernández-Chico C, et al. 1989. Induction of a growth-phase-dependent promoter triggers transcription of *bolA*, an *Escherichia coli* morphogene. EMBO, 12: 3923-3931

Altuvia S, Almirón M, Huisman G, et al. 1994. The *dps* promoter is activated by OxyR during growth and by IHF and sigma S in stationary phase. Molecular Microbiology, 13: 265-272

Aviv M, Giladi H, Schreiber G, et al. 1994. Expression of the genes coding for the *Escherichia coli* integration host factor are controlled by growth phase, rpoS, ppGpp and by autoregulation. Molecular Microbiology, 14: 1021-1031

Azam TA, Iwata A, Nishimura A, et al. 1999. Growth phase-dependent variation in protein composition of the *Escherichia coli* nucleoid. Journal of Bacteriology, 181: 6361-6370

Ball CA, Osuna R, Ferguson KC, et al. 1992. Dramatic changes in Fis levels upon nutrient upshift in *Escherichia coli*. Journal of Bacteriology, 174: 8043–8056

Baba T, Ara T, Hasegawa M, et al. 2006.Construction of *Escherichia coli* K-12 in-frame, single-gene knockout mutants: the Keio collection. Molecular Systems Biology, 2: 0008

Barrett T, Troup DB, et al. 2011. NCBI GEO: archive for functional genomics data sets—10 years on. Nucleic Acids Research, 39: D1005-D1010

Battesti A, Majdalani N, Gottesman S. 2011.The RpoS-mediated general stress response in *Escherichia coli*. Annual Reviews in Microbiology, 65: 189-213

Bykowski T, Sirko A. 1998. Selected phenotypes of *ihf* mutants of *Escherichia coli*. Biochimie, 80: 987-1001

Dillon SC, Dorman CJ. 2010. Bacterial nucleoid-associated proteins, nucleoid structure and gene expression. Nature Reviews Microbiology, 8: 185-195

Falconi M, Brandi A, La Teana A, et al. 1996. Antagonistic involvement of FIS and H-NS proteins in the transcriptional control of *hns* expression. Molecular Microbiology, 19: 965-975

Falconi M, Higgins NP, Spurio R, et al. 1993. Expression of the gene encoding the major bacterial nucleoid protein H-NS is subject to transcriptional auto-repression. Molecular Microbiology, 10: 273-282

Ferullo DJ, Cooper DL, Moore HR, et al. 2009. Cell cycle synchronization of *Escherichia coli* using the stringent response, with fluorescence labeling assays for DNA content and replication. Nature Methods, 41: 8-13

Finkel SE, Johnson RC. 1992. The Fis protein: it's not just for DNA inversion anymore. Molecular Microbiology, 6: 3257-3265

Free A, Dorman CJ. 1995. Coupling of *Escherichia colihns* mRNA levels to DNA synthesis by autoregulation: implications for growth phase control. Molecular Microbiology, 18: 101-113

Giangrossi M, Gualerzi CO, Pon CL. 2001. Mutagenesis of the downstream region of the *Escherichia coli hns* promoter. Biochimie, 83: 251-259

Giangrossi M, Zattoni S, Tramonti A, et al. 2005. Antagonistic role of H-NS and GadX in the regulation of the glutamate decarboxylase-dependent acid resistance system in *Escherichia coli*. The Journal of Biological Chemistry, 280: 21498-21505

Goosen N, van de Putte P. 1995. The regulation of transcription initiation by integration host factor. Molecular Microiology, 16: 1-7

Grainger DC, Goldberg MD, Lee DJ, et al. 2008. Selective repression by Fis and H- NS at the *Escherichia colidps* promoter. Molecular Microbiology, 68: 1366-1377

Harman JG. 2001. Allosteric regulation of the cAMP receptor protein. Biochimica et Biophysica Acta, 1547: 1-17

Hengen PN, Bartram SL, Stewart LE, et al. 1997. Information analysis of Fis binding sites. Nucleic Acids Research, 25: 4994-5002

Hommais F. 2004. GadE (YhiE): a novel activator involved in the response to acid environment in *Escherichiacoli*. Microbiology, 150: 61-72

Hommais F, Krin E, Laurent-WinterC, et al. 2001. Large scaling monitoring of pleiotropic regulation of gene expression by the prokaryotic nucleoid-associated protein, H-NS. Molecular Microbiology, 40: 20-36

Kaufman M, Soule C, Thomas R. 2007. A new necessary condition on interaction graphs for multistationarity. Journal of Theoretical Biology, 248: 675-685

Klumpp S, Zhang Z, Hwa T. 2009. Growth Rate-Dependent Global Effects on Gene Expression in Bacteria.

Cell, 139: 1366-1375

Kolter R, Siegele DA, Tormo A. 1993. The stationary phase of the bacterial life cycle. Annual Reviews in Microbiology, 47: 855-874

Lange R, Hengge-Aronis R. 1991. Identification of a central regulator of stationary-phase gene expression in *Escherichia coli*. Molecular Microbiology, 5: 49-59

Ma Z, Richard H, Tucker DL, et al. 2002.Collaborative regulation of *Escherichia coli* glutamate-dependent acid resistance by two AraC-like regulators, GadX and GadW (YhiW). Journal of Bacteriology, 184: 7001-7012

Martinez-Antonio A. 2011. *Escherichia coli* transcriptional regulatory network. Network Biology, 1: 21-33

Martínez-Antonio A, Lomnitz JG, Sandoval S, et al. 2012. Regulatory design governing progression of population growth phases in bacteria. Plos ONE, 70: e30654

Martínez-Antonio A, Janga SC, Thieffry D. 2008.Functional organisation of *Escherichia coli* transcriptional regulatory network. Journal of Molecular Biology, 38: 238-247

Mechulam Y, Blanquet S, Fayat G. 1987. Dual level control of the *Escherichia colipheST-himA* operon expression.tRNA(Phe)-dependent attenuation and transcriptional operator-repressor control by *himA* and the SOS network. Journal of Molecular Biology, 197: 453-470

Monod J. 1949. The growth of bacterial cultures. Annual Reviews in Microbiology, 3: 371-394

Nasser W, Schneider R, Travers A, et al. 2001. CRP modulates *fis* transcription by alternate formation of activating and repressing nucleoprotein complexes. The Journal of Biological Chemistry, 276: 17878-17886

Nasser W, Rochman M, Muskhelishvili G. 2002. Transcriptional regulation of *fis* operon involves a module of multiple coupled promoters. EMBO, 21: 715-724

Pratt TS, Steiner T, Feldman LS, et al. 1997.Deletion analysis of the *fis* promoter region in *Escherichia coli*: antagonistic effects of integration host factor and Fis. Journal of Bacteriology, 179: 6367-6377

Salgado H, Peralta-Gil M, Gama-Castro et al. 2013. RegulonDB v8. 0: omics data sets, evolutionary conservation, regulatory phrases, cross-validated gold standards and more. Nucleic Acids Research, 41: D203-D213

Sangurdekar DP, Srienc F, Khodursky AB. 2006. A classification based framework for quantitative description of large-scale microarray data. Genome Biology, 7: R32

Semsey S, Jauffred L, Csiszovszki Z, et al. 2013. The effect of LacI autoregulation on the performance of the lactose utilization system in *Escherichia coli*. Nucleic Acids Research, 41: 6381-6390

Shin S, Castanie-Cornet MP, Foster JW, et al. 2001. An activator of glutamate decarboxylase genes regulates the expression of enteropathogenic *Escherichia coli* virulence genes through control of the plasmid-encoded regulator, Per. Molecular Microbiology, 41: 1133-1150

Soares NC, Spät P, Krug K, et al. 2013. Global dynamics of the *Escherichia coli* proteome and phosphoproteome during growth in minimal medium. Journal of Proteome Research, 12: 2611-2621

Taniguchi Y, Choi PJ, Li GW, et al. 2010. Quantifying *E. coli* proteome and transcriptome with single-molecule sensitivity in single cells. Science, 329: 533-538

Thomas R, Thieffry D, Kaufman M.1995. Dynamical behaviour of biological regulatory networks—I. Biological role of feedback loops and practical use of the concept of the loop-characteristic state. Bulletin of Mathematical Biology, 57: 247-276

Tramonti A, De Canio M, De Biase D. 2008. GadX/GadW-dependent regulation of the *Escherichia coli* acid fitness island: transcriptional control at the *gadY-gadW* divergent promoters and identification of four novel 42 bpGadX/GadW-specific binding sites. Molecular Microbiology, 70: 965-982

Tramonti A, Visca P, De Canio M, et al. 2002. Functional characterization and regulation of *gadX*, a gene encoding an AraC/XylS-like transcriptional activator of the *Escherichia coli* glutamic acid decarboxylase system. Journal of Bacteriology, 184: 2603-2613

Ueguchi C, Kakeda M, Mizuno T. 1993.Autoregulatory expression of the *Escherichia coli hns* gene encoding a nucleoid protein: H-NS functions as a repressor of its own transcription. Molecular & General Genetics, 236: 171-178

Valjavec-Gratian M, Henderson TA, Hill TM.2005. Tus-mediated arrest of DNA replication in *Escherichia coli* is modulated by DNA supercoiling. Molecular Microbiology, 58: 758-773

Węgleńska A, Jacob B, Sirko A. 1996. Transcriptional pattern of *Escherichia coli ihfB* (*himD*) gene expression. Gene, 181: 85-88

Wiśniewskia JR, Rakusb D. 2014. Multi-enzyme digestion FASP and the 'Total Protein Approach'-based absolute quantification of the *Escherichia coli* proteome. Journal of Proteomics, 109: 322-331

Yamamoto K, Ishihama A, Busby SJW, et al. 2011. The *Escherichia coli* K-12 MntRminiregulon includes *dps*, which encodes the major stationary-phase DNA-binding protein. Journal of Bacteriology, 193: 1477-1480

Zaslaver A, Bren A, Ronen M, et al. 2006. A comprehensive library of fluorescent transcriptional reporters for *Escherichia coli*. Nature Methods, 3: 623-628

Zheng D, Constantinidou C, Hobman JL, et al. 2004. Identification of the CRP regulon using *in vitro* and *in vivo* transcriptional profiling. Nucleic Acids Research, 32: 5874-5893

Zwietering MH, Jongenburger I, Rombouts FM, et al. 1990. Modeling of the bacterial growth curve. Applied and Environmental Microbiology, 56: 1875-1881

19

Identification of crucial metabolites/reactions in tumor signaling networks

JingRon Li[1], WenJun Zhang[1,2]

[1]School of Life Sciences, Sun Yat-sen University, Guangzhou 510275, China; [2]International Academy of Ecology and Environmental Sciences, Hong Kong

E-mail: zhwj@mail.sysu.edu.cn,wjzhang@iaees.org

Abstract

Changes in metabolites/reactions of cell signaling pathways play a key role in tumorigenesis. In present study, betweenness centrality, degree and k-core value of every metabolite/reaction in tumor signaling pathways p53, AKT, Ras, JAK-STAT, TNF, and VEGF were calculated. Crucial metabolites/reactions in these tumor signaling networks were identified using betweenness centrality. The p53-P-P was identified as the most important metabolite/reaction in p53 signaling pathway, followed by (Ac-p53-P)2 and DNA damage; Akt is the most important metabolite/reaction in AKT signaling pathway, followed by PI3K and PIP3; Ras-GTP is the most important metabolite/reaction in TNF signaling pathway, followed by MEKK1, JNKK and Ras-GDP. The k-core analysis showed that VEGF signaling pathway is the most compact network among these signaling pathways.

Keywords tumor; signaling pathway; crucial metabolite/reaction; k-core analysis; betweenness centrality.

1 Introduction

Changes in metabolites/reactions of cell signaling pathways play a key role in tumorigenesis. A large number of ligands, receptors, signaling proteins and links exist in the signaling pathways of a human cell. Complex signaling pathways form various metabolic networks and affect the metabolic processes of tumors.

Tumor signaling pathways of the human body include mainly JAK-STAT signaling pathway (Marrero, 2005), p53 signaling pathway (Himes et al., 2006; Ho et al, 2006), NF-κB signaling pathway, Ras, PI3K and mTOR signaling pathway (Kolch, 2002; Stauffer et al., 2005), Wnt NF-κB signaling pathway (Katoh, 2005) and BMP signaling pathway (Moustakas, 2002), etc. Various ligands, receptors and signaling proteins are associating with these signaling pathways. They form a complex and directed network, similar to the various networks reported (Ibrahim et al., 2011; Huang and Zhang, 2012; Zhang, 2012a).

Previous studies of tumor signaling pathways focus mainly on the metabolic processes and chemical processes of some selected metabolites, and the tumorigenesis induced by abnormal signaling from mutation of this metabolite or gene. Detailed studies were also conducted on the chemical structure and the metabolic pathways of ligands, receptors, and signaling proteins. A most recent study on network analysis of tumor signaling networks was the degree distribution analysis of tumor signaling networks (Huang and Zhang, 2012). In their study, Huang and Zhang (2012) considered that the metabolites/reactions (steps) with higher degree are often crucial metabolites/reactions.

Although a tumor signaling pathway may be very complex, some metabolites/reactions in the network are more important for tumorigenesis than the remaining metabolites/reactions. Using network analysis methods, the present study will identify crucial metabolites/reactions of some important tumor signaling pathways, aiming to provide valuable clues for further studies.

2 Material and Methods

2.1 Data sources

Six tumor signaling pathways are closely related to various metabolites/reactions (Huang and Zhang, 2012). The later were collated and interpreted to form a new analytical database. All image information for signaling networks were downloaded from SABiosciences (http://www.sabiosciences.com/pathwaycentral.php) (Pathway Central, 2012) and Abcam (http://www.abcam.com/) (Abcam, 2012).

2.2 Data conversion

Each metabolite/reaction (node) in the tumor signaling image was given an ID number to generate a network. Open Data/Data editors/Matrix editor in the UCINET software, and then paste the network data into the matrix editor. After that we saved the network data as a file in ".##h" format. In the NetDraw program, we choose the File/Open/Ucinet dataset/Network and open the ".##h" file, and then choose File/Save data as/Pajek/Net file to save it as a file in ".net" format (Kuang and Zhang, 2011). Through this step, all tumor signaling images were interpreted as the network data used in Pajek.

2.3 Software and methods

2.3.1 UCINET

The UCINET software on network analysis integrated the NetDraw, a program for one- and two-dimensional network analysis, and the ongoing three-dimensional display application, Mage, etc. It also integrated some application programs of Pajek that used in large-scale network analysis. The UCINET software can read text files, KrackPlot, Pajek, Negopy, and VNA files. It can handle the networks as large as 32767 nodes.

2.3.2 Pajek

Pajek is the large and complex software for network analysis, which is characterized by quick computing, higher degree of visualization and abstraction. It can handle the networks with millions of nodes. Pajek provides abstract methods for handling complex networks, and is mainly used in analyzing the global structure of networks.

Degree

Degree is the most basic property of a complex network (Kuang and Zhang, 2011; Huang and Zhang, 2012; Zhang, 2012a, b). The degree of a node is defined as the number of its connected nodes. In a sense, the larger the degree of a node, the more important the node is (Zhang, 2012a, b). In a directed network, the degree is the sum of incoming degree and outgoing degree. Choosing In/Out/All commands of Net/Partitions/Degree menu in Pajek, the degree, incoming degree and outgoing degree will be calculated.

Betweenness centrality

Betweenness centrality is the measure of a node's centrality in a network (Zhang, 2012a, b). It is equal to the number of shortest paths from all nodes to all others that pass through that node. In general, betweenness centrality is a more useful measure than degree in identifying the importance of a node. For a network, betweenness centrality is more global than degree.

k-neighbor and k-core network

If the node i and node j are connected with an edge, then the two nodes are neighbors. If the node i connects to node j by going through k edges, then two nodes are called k-nearest neighbor (k-neighbor) for each other. For a directed graph, there are two types of k-nearest neighbors, k-out and -in nearest neighbors. Starting from the node i, if there are k positive-directed edges between nodes i and j, and j is called the k-out nearest neighbor of i and, i is called the k-in nearest neighbor of j. For a directed graph, choose Net/k-Neighbors/Output in Pajek, and input the longest distance k value in the pop-up dialog box (k=0 means finding all k-neighbors of the node). The result is a partition file, in which the ID number of the class that node i belongs to is the shortest path from required node to node i. In addition, choosing Net/k-Neighbors/All in Pajek, will output the k-neighbors of nodes for an undirected graph.

In a network, if any node has at least k neighbors that belong to the network, the network is called a k-core network. To find the core of a complex network means finding all k-core networks in the complex network. A large total k-core value means a compact network.

3 Results

3.1 p53 signaling pathway

The image of p53 signaling pathway is shown in Fig. 1.

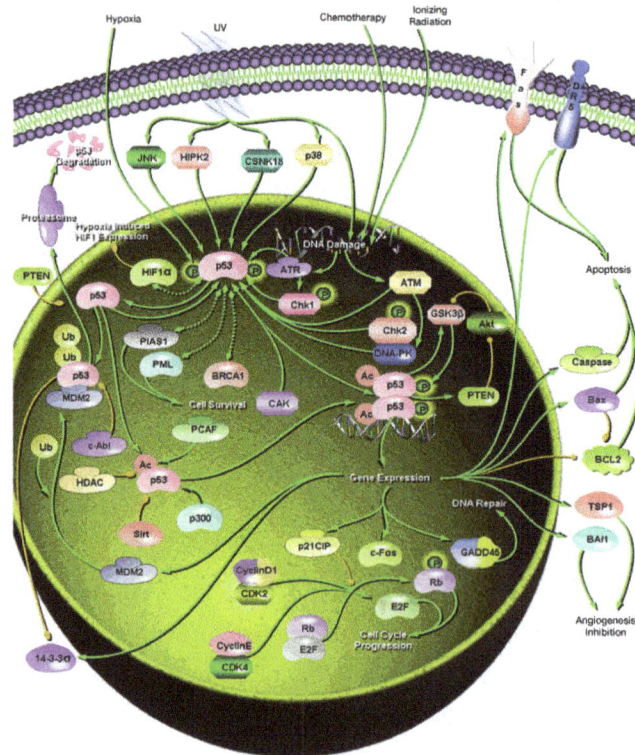

Fig. 1 P53 signaling pathway (Pathway Central, 2012)

Using Netdraw, the network graph of P53 signaling pathway in Fig. 1 was drawn, as indicated in Fig. 2.

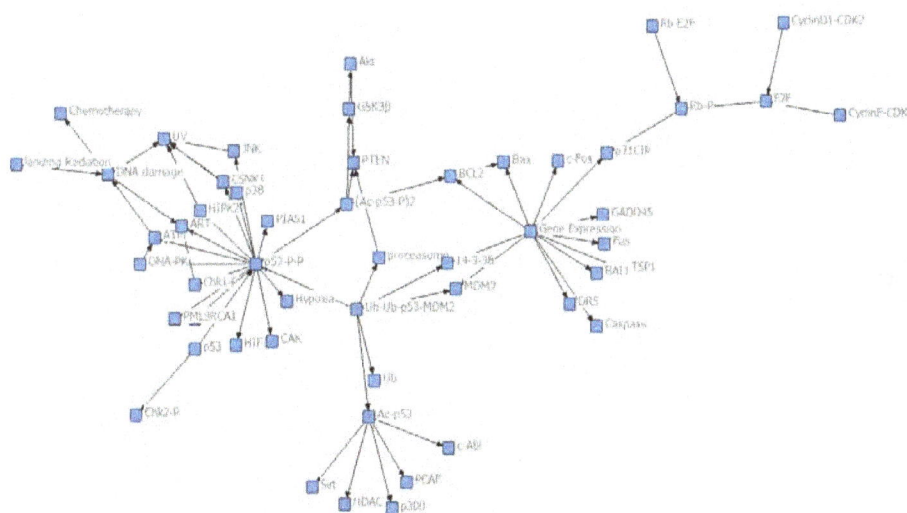

Fig. 2 The network graph of p53 signaling pathway.

Table 1 Degree of metabolites/reactions in p53 signaling pathway.

Metabolites/ reactions	Degree	Metabolites/ reactions	Degree	Metabolites/ reactions	Degree	Metabolites/ reactions	Degree
Hypoxia	1	p38	2	HDAC	1	E2F	3
UV	5	BRCA1	1	DNA-PK	2	Rb-E2F	1
Chemotherapy	1	PTEN	3	Akt	2	CyclinD1-CDK2	1
ATM	3	PML	1	Chk2-P	1	CyclinE-CDK4	1
DNA damage	5	proceasome	2	GSK3β	2	BAI1	1
Ionizing Radiation	1	PIAS1	1	p21CIP	2	Fas	1
ART	3	c-Abl	1	Caspase	1	DR5	1
JNK	2	PCAF	1	c-Fos	1	Gene Expression	12
Chk1-P	2	Ub	1	Bax	2	(Ac-p53-P)2	5
HIPK2	2	p300	1	GADD45	1	Ac-p53	6
HIFα	1	MDM2	2	BCL2	3	Ub-Ub-p53-MDM2	6
CSNK1	2	Sirt	1	Rb-P	3	p53	2
CAK	1	14-3-3θ	2	TSP1	1	p53-P-P	18

Counted degree of metabolites/reactions is listed in Table 1.

Use Pajek/Net/Partitions/Core/All to calculate all k-core networks of p53 signaling pathway. The result is a partition file, in which the ID number of the class that node i belongs to is the largest k value of all k-core networks of node i. The results showed that k-core value of UV, ATM, DNA damage, ART, JNK, Chk1-P, HIPK2, CSNK1, p38, PTEN, proceasome, MDM2, 14-3-3θ, DNA-PK, Akt, GSK3β, Bax, BCL2 Gene, Expression, (Ac-p53-P)2, Ub-Ub-p53-MDM2, and p53-P-P is 2, and k-core value is 1 for remaining metabolites/reactions.

Use Pajek/Vector/Centrality/Betweeness to calculate the betweenness centrality of p53 signaling pathway.

The results are listed in Table 2. A larger betweenness centrality means that the corresponding metabolite/reaction is more crucial in the signaling pathway. Thus, p53-P-P is identified as the most important metabolite/reaction in p53 signaling pathway, followed by (Ac-p53-P)2, DNA damage, and ATM.

Table 2 Betweenness centrality (BC) of metabolites/reactions in p53 signaling pathway.

Metabolites/ reactions	BC	Metabolites/ reactions	BC	Metabolites/ reactions	BC	Metabolites/ reactions	BC
Hypoxia	0	p38	0.000392	HDAC	0	E2F	0.001569
UV	0	BRCA1	0	DNA-PK	0	Rb-E2F	0
Chemotherapy	0	PTEN	0	Akt	0.000392	CyclinD1-CDK2	0
ATM	0.004706	PML	0	Chk2-P	0	CyclinE-CDK4	0
DNA damage	0.005098	proceasome	0.000392	GSK3β	0.001569	BAI1	0
Ionizing Radiation	0	PIAS1	0	p21CIP	0	Fas	0
ART	0	c-Abl	0	Caspase	0	DR5	0
JNK	0.000392	PCAF	0	c-Fos	0	Gene Expression	0
Chk1-P	0	Ub	0	Bax	0	(Ac-p53-P)2	0.00549
HIPK2	0.000392	p300	0	GADD45	0	Ac-p53	0.001961
HIFα	0	MDM2	0	BCL2	0.001569	Ub-Ub-p53-MDM2	0
CSNK1	0.000392	Sirt	0	Rb-P	0.001569	p53	0
CAK	0	14-3-3θ	0	TSP1	0	p53-P-P	0.024314

3.2 AKT signaling pathway

Fig. 3 is the network graph of AKT signaling pathway.

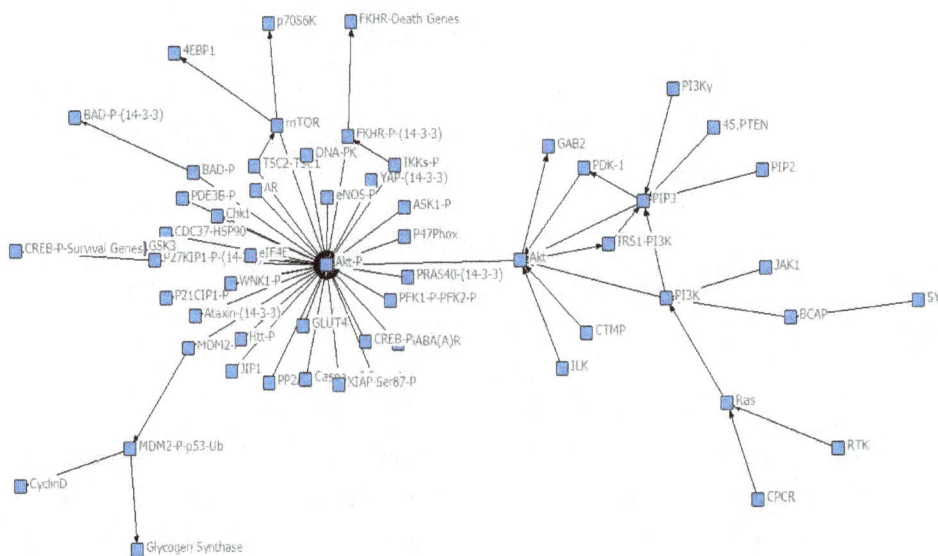

Fig. 3 The network graph of AKT signaling pathway.

Table 3 Degree of metabolites/reactions in AKT signaling pathway.

Metabolites/ reactions	Degree	Metabolites/ reactions	Degree	Metabolites/ reactions	Degree	Metabolites/ reactions	Degree
GABA(A)R	1	Raf1	1	GSK3	1	JAK1	1
CPCR	1	Caspase9-P	1	CREB-P-Survival Genes	1	BCAP	1
Ras	3	XIAP-Ser87-P	1	CyclinD	1	PI3Ky	2
PI3K	5	PDE3B-P	1	FKHR-P-(14-3-3)	1	GLUT4	1
GAB2	1	BAD-P	2	Glycogen Synthase	3	Akt	1
RTK	1	TSC2-TSC1	2	FKHR-Death Genes	1	Akt-P	8
ILK	1	BAD-P-(14-3-3)	1	PFK1-P-PFK2-P	1	JIP1	33
PIP2	1	mTOR	4	YAP-(14-3-3)	1	ASK1-P	1
SYK	1	Chk1	1	IKKs-P	1	4EBP1	1
IRS1-PI3K	2	p70S6K	1	Ataxin-(14-3-3)	2	eIF4E	1
CTMP	1	P21CIP1-P	1	Htt-P	1	AR	1
PIP3	7	MDM2-P	2	P47Phox	1	DNA-PK	1
PP2A	1	P27KIP1-P-(14-3-3)	2	WNK1-P	1	eNOS-P	1
PDK-1	2	MDM2-P-p53-Ub	3	PRAS40-(14-3-3)	1		
CDC37-HSP90	1	CREB-P	3	45.PTEN	1		

Table 4 Betweenness centrality (BC) of metabolites/reactions in AKT signaling pathway.

Metabolites/ reactions	BC	Metabolites/ reactions	BC	Metabolites/ reactions	BC	Metabolites/ reactions	BC
GABA(A)R	0	Raf1	0	GSK3	0	JAK1	0
CPCR	0	Caspase9-P	0	CREB-P-Survival Genes	0	BCAP	0.002193
Ras	0.004386	XIAP-Ser87-P	0	CyclinD	0	PI3Ky	0
PI3K	0.01127	PDE3B-P	0	FKHR-P-(14-3-3)	0.000313	GLUT4	0
GAB2	0	BAD-P	0	Glycogen Synthase	0	Akt	0.015351
RTK	0	TSC2-TSC1	0	FKHR-Death Genes	0	Akt-P	0
ILK	0	BAD-P-(14-3-3)	0	PFK1-P-PFK2-P	0	JIP1	0
PIP2	0	mTOR	0.000627	YAP-(14-3-3)	0	ASK1-P	0
SYK	0	Chk1	0	IKKs-P	0	4EBP1	0
IRS1-PI3K	0.00219	p70S6K	0	Ataxin-(14-3-3)	0	eIF4E	0
CTMP	0	P21CIP1-P	0	Htt-P	0	AR	0
PIP3	0.00908	MDM2-P	0	P47Phox	0	DNA-PK	0
PP2A	0	P27KIP1-P-(14-3-3)	0	WNK1-P	0	eNOS-P	0
PDK-1	0	MDM2-P-p53-Ub	0.000627	PRAS40-(14-3-3)	0		
CDC37-HSP90	0	CREB-P	0	45.PTEN	0		

Counted degree of metabolites/reactions is listed in Table 3.

The results showed that k-core value of FKHR-P-(14-3-3), IKKs-P, Akt, Akt-P, mTOR, TSC2-TSC1, PDK-1, PIP3, IRS1-PI3K, and PI3K is 2, and k-core value is 1 for remaining metabolites/reactions.

Betweenness centrality of metabolites/reactions showed that Akt is the most important metabolite/reaction in AKT signaling pathway, followed by PI3K and PIP3 (Table 4).

3.3 JAK-STAT signaling pathway

Fig. 4 is the network graph of JAK-STAT signaling pathway.

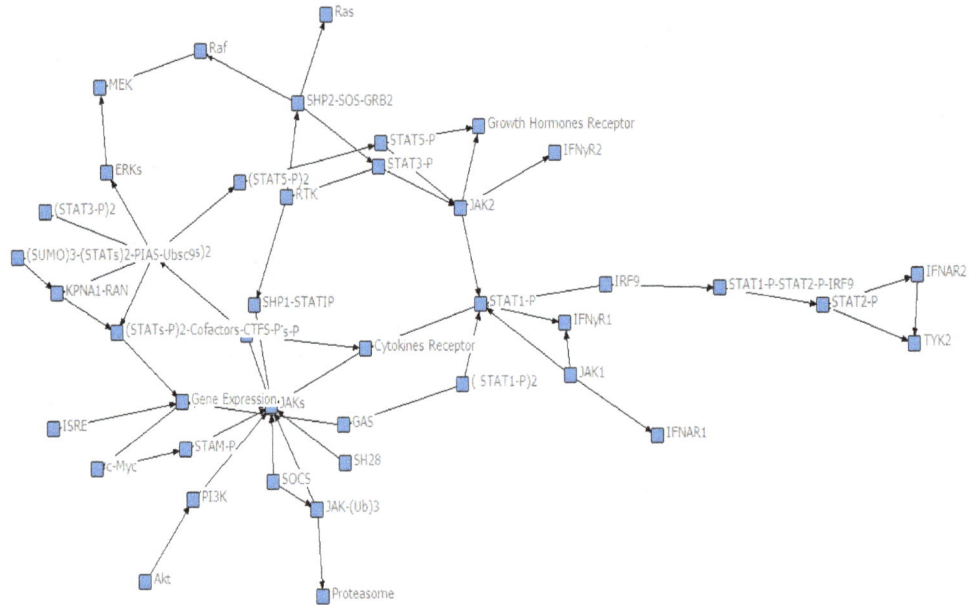

Fig. 4 The network graph of JAK-STAT signaling pathway.

Table 5 Degree of metabolites/reactions in JAK-STAT signaling pathway.

Metabolites/ reactions	Degree	Metabolites/ reactions	Degree	Metabolites/ reactions	Degree	Metabolites/ reactions	Degree
JAK1	3	RTK	3	MEK	2	STAT1-P-STAT2-P-IRF9	2
JAK2	5	IFNAR2	2	PI3K	2	(SUMO)3-(STATs)2-PIAS-Ubsc9	2
TYK2	2	STAT3-P	3	ERKs	2	KPNA1-RAN	3
JAKs	9	STAT1-P	6	Akt	1	GAS	2
SHP1-STATIP	2	STAT2-P	3	STAM-P	2	ISRE	1
Growth Hormones	2	STATs-P	3	c-Myc	2	(STATs-P)2-Cofactors-CTFS-P	3
IFNyR1	2	SH28	1	(STAT5-P)2	2	JAK-(Ub)3	3
STAT5-P	3	SOCS	2	(STAT3-P)2	1	Proteasome	1
IFNyR2	1	SHP2-SOS-GRB2	4	(STAT1-P)2	2	Gene Expression	5
Cytokines Receptor	3	Ras	1	(STATs)2	7		
IFNAR1	1	Raf	2	IRF9	2		

Counted degree of metabolites/reactions is listed in Table 5.

The results showed that k-core value of IFNyR2, IFNAR1, SH28, Ras, PI3K, Akt, (STAT3-P)2, ISRE, and Proteasome is 1, and k-core value is 2 for remaining metabolites/reactions.

Betweenness centrality of metabolites/reactions showed that JAK2 is the most important metabolite/reaction in JAK-STAT signaling pathway, followed by (STARTs)2, Gene Expression, START5-P and (START5-P)2 (Table 6).

Table 6 Betweenness centrality (BC) of metabolites/reactions in JAK-STAT signaling pathway.

Metabolites/ reactions	BC	Metabolites/ reactions	BC	Metabolites/ reactions	BC	Metabolites/ reactions	BC
JAK1	0	RTK	0.004878	MEK	0	STAT1-P-STAT2-P-IRF9	0.001829
JAK2	0.021951	IFNAR2	0	PI3K	0.00061	(SUMO)3-(STATs)2-PIAS-Ubsc9	0
TYK2	0	STAT3-P	0.009146	ERKs	0.001829	KPNA1-RAN	0.003049
JAKs	0	STAT1-P	0.016768	Akt	0	GAS	0.002134
SHP1-STATIP	0.001829	STAT2-P	0.002439	STAM-P	0.00061	ISRE	0
Growth Hormones	0	STATs-P	0	c-Myc	0.006098	(STATs-P)2-Cofactors-CTFS-P	0.011585
IFNyR1	0	SH28	0	(STAT5-P)2	0.012195	JAK-(Ub)3	0.00061
STAT5-P	0.014634	SOCS	0	(STAT3-P)2	0	Proteasome	0
IFNyR2	0	SHP2-SOS-GRB2	0.007927	(STAT1-P)2	0	Gene Expression	0.015549
Cytokines Receptor	0.003963	Ras	0	(STATs)2	0.019512		
IFNAR1	0	Raf	0.001829	IRF9	0		

3.4 TNF signaling pathway

Fig. 5 is the network graph of TNF signaling pathway.

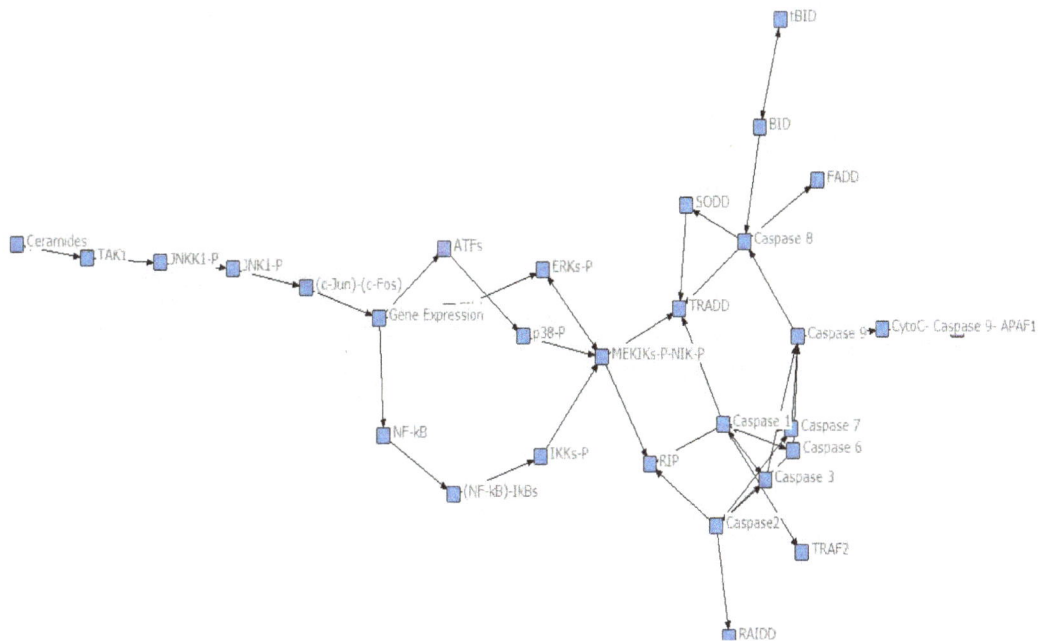

Fig. 5 The network graph of TNF signaling pathway.

Table 7 Degree of metabolites/reactions in TNF signaling pathway.

Metabolites/ reactions	Degree	Metabolites/ reactions	Degree	Metabolites/ reactions	Degree	Metabolites/ reactions	Degree
TRADD	4	Caspase 6	4	Cytoc	2	EIk1	2
SODD	2	Caspase 2	2	IKKs-P	1	Ceramides	1
TRAF2	1	Caspase 7	1	CytoC- Caspase 9- APAF1	2	JNK1-P	2
FADD	1	Caspase 1	2	p38-P	2	JNKK1-P	2
RIP	3	BID	2	(NF-kB)-IkBs	1	TAK1	2
RAIDD	1	Caspase 9	2	ATFs	2	(c-Jun)-(c-Fos)	1
Caspase 3	4	tBID	2	NF-kB	2	Gene Expression	2
Caspase 8	5	MEKIKs-P-NIK-P	2	ERKs-P	1		

Table 8 Betweenness centrality (BC) of metabolites/reactions in TNF signaling pathway.

Metabolites/ reactions	BC	Metabolites/ reactions	BC	Metabolites/ reactions	BC	Metabolites/ Reactions	BC
TRADD	0	Caspase 6	0.009195	Cytoc	0	EIk1	0.017241
SODD	0	Caspase 2	0.006322	IKKs-P	0.011494	Ceramides	0
TRAF2	0	Caspase 7	0.008046	CytoC- Caspase 9-	0.006897	JNK1-P	0.041379
FADD	0	Caspase 1	0.023563	p38-P	0.016092	JNKK1-P	0.029885
RIP	0	BID	0.004598	(NF-kB)-IkBs	0.013793	TAK1	0.016092
RAIDD	0	Caspase 9	0.028736	ATFs	0.017241	(c-Jun)-(c-Fos)	0.050575
Caspase 3	0.008046	tBID	0	NF-kB	0.013793	Gene Expression	0.057471
Caspase 8	0.021839	MEKIKs-P-NIK-P	0.041379	ERKs-P	0.02069		

Counted degree of metabolites/reactions is listed in Table 7.

The results showed that k-core value of Caspase 3, Caspase 2, Caspase 6, Caspase 7, Caspase 1, and Caspase 9 is 3, and k-core value is 2 or 1 for remaining metabolites/reactions.

Betweenness centrality of metabolites/reactions showed that Gene Expression is the most important metabolite/reaction in TNF signaling pathway, followed by (c-Jun)-(c-Fos), MEKIKs-P-NIK-P and JNK1-P (Table 8).

3.5 Ras signaling pathway

Counted degree of metabolites/reactions is listed in Table 9.

The results showed that k-core value of Actin Cytoskeleton, PMA, TCR, and Rho is 1, and k-core value is 2 for remaining metabolites/reactions.

Betweenness centrality of metabolites/reactions showed that Ras-GTP is the most important metabolite/reaction in Ras signaling pathway, followed by MEKK1, JNKK and Ras-GDP (Table 10).

Table 9 Degree of metabolites/reactions in Ras signaling pathway.

Metabolites/ reactions	Degree	Metabolites/ reactions	Degree	Metabolites/ reactions	Degree	Metabolites/ reactions	Degree
Integrins	2	RalGDS	3	RalBP1	3	MEKK1	3
Rap1A-GTP	4	MEKs-P	3	PMA	1	Rho	1
PLC-$\sum$	4	Ral	4	CD-GECII	2	p190-B	2
Ras-GDP	7	ERKs-P	3	PI3K	3	JNKK	2
Ras-GTP	16	PLD	2	Rac	5	JNK	3
GRB2	2	ERKs	4	TCR	1	c-Jun-c-Fun	2
GAP	2	CDC42	3	Lck	2	ATF2	2
GEF	2	Elk1	3	PAKs	2	Gene Expression	3
Raf-P	4	ActinCytoskeleton	1	p120-GAP	2		

Table 10 Betweenness centrality (BC) of metabolites/reactions in Ras signaling pathway.

Metabolites/ reactions	BC	Metabolites/ reactions	BC	Metabolites/ reactions	BC	Metabolites/ Reactions	BC
Integrins	0	RalGDS	0.042781	RalBP1	0.003565	MEKK1	0.108734
Rap1A-GTP	0.035651	MEKs-P	0.061497	PMA	0	Rho	0
PLC-$\sum$	0.067736	Ral	0.024955	CD-GECII	0.050802	p190-B	0.008021
Ras-GDP	0.088235	ERKs-P	0.042781	PI3K	0.02139	JNKK	0.090018
Ras-GTP	0.348485	PLD	0	Rac	0.062389	JNK	0.069519
GRB2	0	ERKs	0.024064	TCR	0	c-Jun-c-Fun	0.011141
GAP	0	CDC42	0.002674	Lck	0	ATF2	0.011141
GEF	0	Elk1	0.001783	PAKs	0.026738	Gene Expression	0
Raf-P	0.0918	ActinCytoskeleton	0	p120-GAP	0.016934		

3.6 VEGF signaling pathway (Matsumoto and Claesson-Welsh, 2001)

Counted degree of metabolites/reactions is listed in Table 11.

The results showed that k-core value of all metabolites/reactions is 4.

Betweenness centrality of metabolites/reactions showed that ANGIO GENESIS is the most important metabolite/reaction in VEGF signaling pathway, followed by PIP3 and Actin Reorganization (Table 12).

Table 11 Degree of metabolites/reactions in VEGF signaling pathway.

Metabolites/ reactions	Degree	Metabolites/ reactions	Degree	Metabolites/ reactions	Degree	Metabolites/ reactions	Degree
Src	4	Akt/PKB	8	Cell Migration	6	Ca++	4
PIP2	6	p38	4	Cell Suvival	6	ANGIO GENESIS	12
VEGFR2	6	P	4	Ras	4	cPLA	4
PLCy-P	4	MAPKAPK2/3	4	NO production	4	Actin Reorganization	7
MKK3/6	4	BAD-P	6	Raf1	6	Gene Expression	5

PI3K-P	6	HSP27	4	MEK1/2	4	Prostagiandin Production	6
FAK-Paxillin	6	Caspase9-P	6	ERK1/2	6	Vascular Cell Permeability	4
PIP3	6	Focal Adhesion Turnover	4	PKC	4	DAG	6
GRB2-SHC-SOS	4	eNOS-HSP90	4	IP3	6		

Table 12 Betweenness centrality (BC) of metabolites/reactions in VEGF signaling pathway.

Metabolites/ reactions	BC	Metabolites/ reactions	BC	Metabolites/ reactions	BC	Metabolites/ Reactions	BC
Src	0.06634	Akt/PKB	0.108081	Cell Migration	0.106046	Ca++	0.044016
PIP2	0.157028	p38	0.028818	Cell Suvival	0.101791	ANGIO GENESIS	0.336376
VEGFR2	0.120395	P	0.000594	Ras	0.063651	cPLA	0.026292
PLCy-P	0.012032	MAPKAPK2/3	0.055853	NO production	0.035001	Actin Reorganization	0.189037
MKK3/6	0.033571	BAD-P	0.039279	Raf1	0.118018	Gene Expression	0.111727
PI3K-P	0.164617	HSP27	0.091949	MEK1/2	0.07546	Prostagiandin Production	0.085258
FAK-Paxillin	0.11937	Caspase9-P	0.039279	ERK1/2	0.120936	Vascular Cell Permeability	0.081735
PIP3	0.208781	Focal Adhesion	0.069296	PKC	0.065359	DAG	0.094058
GRB2-SHC-SOS	0.060977	eNOS-HSP90	0.017473	IP3	0.067367		

4 Discussion

The results showed that the degree-based relative importance of metabolite/reaction is somewhat different from the betweenness centrality-based relative importance. We adopted the later results. Compared to other signaling pathways, all metabolites/reactions in VEGF signaling pathway have a k-core value as high as 4. This means that all metabolites/reactions in VEGF signaling pathway are closely connected.

In this study we did not conduct analysis on undirected networks. Directed networks should be analyzed and more methods and tools should be used to approach tumor signaling pathways in the future (Zhang, 2012a, b).

Acknowledgment

The authors are thankful to Ms LQ Jiang and Na Li for their help in article preparation. Two authors contributed the same to the article.

References

Abcam. 2012. http://www.abcam.com/. Abcam, USA

Himes SR, Sester DP, Ravasi T, et al. 2006. The JNK are important for development and survival of macrophages. Journal of Immunology, 176(4): 2219-2228

Ho CC, Siu WY, Lau A, Chan WM, et al. 2006. Stalled replication induces p53 accumulation through distinct mechanisms from DNA damage checkpoint pathways. Cancer Research, 66(4): 2233-2241

Huang JQ, Zhang WJ. 2012. Analysis on degree distribution of tumor signaling networks. Network Biology,

2(3): 95-109

Ibrahim SS, Eldeeb MAR, Rady MAH. 2011. The role of protein interaction domains in the human cancer network. Network Biology, 1(1): 59-71

Katoh M. 2005. WNT/PCP signaling pathway and human cancer. Oncology Reports, 14(6): 1583-1588

Kolch W. 2002. Ras/Raf signalling and emerging pharmacotherapeutic targets. Expert Opinion Pharmaco-therapy, 3(6): 709-718

Kuang WP, Zhang WJ. 2011. Some effects of parasitism on food web structure: a topological analysis. Network Biology, 1(3-4): 171-185

Marrero MB. 2005. Introduction to JAK/STAT signaling and the vasculature.Vascul Pharmacology, 43(5): 307-309

Matsumoto T, Claesson-Welsh L. 2001. VEGF receptor signal transduction. Sci STKE. 112: RE21

Moustakas A, Pardali K, Gaal A, et al. 2002. Mechanisms of TGF-beta signaling in regulation of cell growth and differentiation. Immunology Letters, 82(1-2): 85-91

Pathway Central. 2012. http://www.sabiosciences.com/pathwaycentral.php. SABiosciences, QIAGEN, USA

Stauffer F, Holzer P, Garcia-Echeverria CB. 2005. Locking the PI3K/PKB pathway in tumor cells. Current Medicinal Chemistry - Anti-Cancer Agents, 5(5): 449-462

Zhang WJ. 2012a. Computational Ecology: Graphs, Networks and Agent-based Modeling. World Scientific, Singapore

Zhang WJ. 2012b. Several mathematical methods for identifying crucial nodes in networks. Network Biology, 2(4): 121-126

Permissions

All chapters in this book were first published in NB, by International Academy of Ecology and Environmental Sciences; hereby published with permission under the Creative Commons Attribution License or equivalent. Every chapter published in this book has been scrutinized by our experts. Their significance has been extensively debated. The topics covered herein carry significant findings which will fuel the growth of the discipline. They may even be implemented as practical applications or may be referred to as a beginning point for another development.

The contributors of this book come from diverse backgrounds, making this book a truly international effort. This book will bring forth new frontiers with its revolutionizing research information and detailed analysis of the nascent developments around the world.

We would like to thank all the contributing authors for lending their expertise to make the book truly unique. They have played a crucial role in the development of this book. Without their invaluable contributions this book wouldn't have been possible. They have made vital efforts to compile up to date information on the varied aspects of this subject to make this book a valuable addition to the collection of many professionals and students.

This book was conceptualized with the vision of imparting up-to-date information and advanced data in this field. To ensure the same, a matchless editorial board was set up. Every individual on the board went through rigorous rounds of assessment to prove their worth. After which they invested a large part of their time researching and compiling the most relevant data for our readers.

The editorial board has been involved in producing this book since its inception. They have spent rigorous hours researching and exploring the diverse topics which have resulted in the successful publishing of this book. They have passed on their knowledge of decades through this book. To expedite this challenging task, the publisher supported the team at every step. A small team of assistant editors was also appointed to further simplify the editing procedure and attain best results for the readers.

Apart from the editorial board, the designing team has also invested a significant amount of their time in understanding the subject and creating the most relevant covers. They scrutinized every image to scout for the most suitable representation of the subject and create an appropriate cover for the book.

The publishing team has been an ardent support to the editorial, designing and production team. Their endless efforts to recruit the best for this project, has resulted in the accomplishment of this book. They are a veteran in the field of academics and their pool of knowledge is as vast as their experience in printing. Their expertise and guidance has proved useful at every step. Their uncompromising quality standards have made this book an exceptional effort. Their encouragement from time to time has been an inspiration for everyone.

The publisher and the editorial board hope that this book will prove to be a valuable piece of knowledge for researchers, students, practitioners and scholars across the globe.

List of Contributors

Amar Ćemanović, Jasmin Šutković and Mohamed Ragab Abdel Gawwad
Genetics and Bioengineering department, International University of Sarajevo, Ilidza, 71220 Bosnia and Herzegovina

Rabab Elamawi and Waleed Elkhoby
Rice Research and Training Center, Sakha, Egypt

Alessandro Ferrarini
Department of Evolutionary and Functional Biology, University of Parma, Via G. Saragat 4, I-43100 Parma, Italy

Mouna Choura and Faiçal Brini
Biotechnology and Plant Improvement Laboratory, Centre of Biotechnology of Sfax (CBS)/University of Sfax, B.P "1177" 3018, Sfax Tunisia

Nahida Habib, Iffat Jabin and Mohammad Motiur Rahman
Department of Computer Science and Engineering (CSE), Mawlana Bhashani Science and Technology University (MBSTU), Santosh, Tangail-1902, Bangladesh

Kawsar Ahmed
Department of Information and Communication Technology (ICT), Mawlana Bhashani Science and Technology University (MBSTU), Santosh, Tangail-1902, Bangladesh
Group of Bio-photomatiχ, Santosh, Tangail-1902, Bangladesh

Alessandro Ferrarini
Department of Evolutionary and Functional Biology, University of Parma, Via G. Saragat 4, I-43100 Parma, Italy

Daniel Andrés Dos Santos
Instituto de Biodiversidad Neotropical, Facultad de Ciencias Naturales e Instituto Miguel Lillo, Universidad Nacional de Tucumán – CONICET. Horco Molle S/N, Yerba Buena, Tucumán, Argentina

Virginia Abdala
Instituto de Biodiversidad Neotropical, Facultad de Ciencias Naturales e Instituto Miguel Lillo, Universidad Nacional de Tucumán – CONICET. Horco Molle S/N, Yerba Buena, Tucumán, Argentina

Instituto de Herpetología, Fundación Miguel Lillo-CONICET. Miguel Lillo 251, San Miguel de Tucumán, Tucumán, Argentina
Cátedra de Biología General, Facultad de Ciencias Naturales e IML, Universidad Nacional de Tucumán, Miguel Lillo 251, San Miguel de Tucumán, Tucumán, Argentina

María Laura Ponssa and María José Tulli
Instituto de Herpetología, Fundación Miguel Lillo-CONICET. Miguel Lillo 251, San Miguel de Tucumán, Tucumán, Argentina

Lavinija Mataković
University of Josip Jurja Strossmayer, Biology Department, Croatia

Lizhi Zhang
Department of Molecular Genetics, The Ohio State University, 484 West 12th Avenue, Columbus, OH 43210, USA

Shruti Jain
Department of Electronics and Communication Engineering, Jaypee University of Information Technology, Waknaghat, Solan, Himachal Pradesh 173234, India

D. S. Chauhan
GLA University, Mathura, Uttar Pradesh, 281406, India

Soumya Banerjee
University of Oxford, Oxford, United Kingdom; 2Ronin Institute, Montclair, USA

Alessandro Ferrarini
Department of Evolutionary and Functional Biology, University of Parma, Via G. Saragat 4, I-43100 Parma, Italy

Yue Zhao
Sun Yat-sen University, Guangzhou 510275, China

WenJun Zhang
Sun Yat-sen University, Guangzhou 510275, China
International Academy of Ecology and Environmental Sciences, Hong Kong
School of Life Sciences, Sun Yat-sen University, Guangzhou 510275, China; International Academy of Ecology and Environmental Sciences, Hong Kong

Priyanka Narad
Amity Institute of Biotechnology, Amity University, Uttar Pradesh, India

Kailash C. Upadhyaya
Amity Institute of Molecular Biology & Genomics, Amity University, Uttar Pradesh, India

Anup Som
Centre of Bioinformatics, Institute of Interdisciplinary Studies, University of Allahabad, Allahabad, India

Mouna Choura
Laboratory of Plant Protection and Improvement, Center of Biotechnology of Sfax, University of Sfax, Route Sidi Mansour Km 6, 3018 Sfax, Tunisia

Ahmed Rebaï
Molecular and Cellular Diagnosis Processes, Center of Biotechnology of Sfax, University of Sfax, Route Sidi Mansour Km 6, 3018 Sfax, Tunisia

Khaled Masmoudi
International Center for Biosaline Agriculture (ICBA), Dubai, United Arab Emirates

Q. Din
Department of Mathematics, University of Poonch Rawalakot, Pakistan

LiQin Jiang
School of Life Sciences, Sun Yat-sen University, Guangzhou 510275, China

Xin Li
College of Plant Protection, Northwest A & F University, Yangling 712100, China; Yangling Institute of Modern Agricultural Standardization, Yangling 712100, China

Alessandro Ferrarini
Department of Evolutionary and Functional Biology, University of Parma, Via G. Saragat 4, I-43100 Parma, Italy

Agustino Martínez-Antonio
Departamento de Ingeniería Genética, Centro de Investigación y de Estudios Avanzados del Instituto Politécnico Nacional, Unidad Irapuato. Km. 9.6 Libramiento Norte Carretera Irapuato-León. CP 36821. Irapuato, Guanajuato, México

Index

www.ingramcontent.com/pod-product-compliance
Lightning Source LLC
Chambersburg PA
CBHW080257230326

41458CB00097B/5087